MATHEMATICAL MORSELS

By
ROSS HONSBERGER

THE
DOLCIANI MATHEMATICAL EXPOSITIONS

Published by
THE MATHEMATICAL ASSOCIATION OF AMERICA

The Dolciani Mathematical Expositions

NUMBER THREE

MATHEMATICAL MORSELS

By
ROSS HONSBERGER
University of Waterloo

Published and Distributed by
THE MATHEMATICAL ASSOCIATION OF AMERICA

Complete Set ISBN 0-88385-300-0
Vol. 3 ISBN 0-88385-303-5

Printed in the United States of America

Current printing (last digit):
10 9 8 7 6 5 4 3 2 1

The DOLCIANI MATHEMATICAL EXPOSITIONS series of the Mathematical Association of America was established through a generous gift to the Association from Mary P. Dolciani, Professor of Mathematics at Hunter College of the City University of New York. In making the gift, Professor Dolciani, herself an exceptionally talented and successful expositor of mathematics, had the purpose of furthering the ideal of excellence in mathematical exposition.

The Association, for its part, was delighted to accept the gracious gesture initiating the revolving fund for this series from one who has served the Association with distinction, both as a member of the Committee on Publications and as a member of the Board of Governors. It was with genuine pleasure that the Board chose to name the series in her honor.

The books in the series are selected for their lucid expository style and stimulating mathematical content. Typically, they contain an ample supply of exercises, many with accompanying solutions. They are intended to be sufficiently elementary for the undergraduate and even the mathematically inclined high-school student to understand and enjoy, but also to be interesting and sometimes challenging to the more advanced mathematician.

———

The following DOLCIANI MATHEMATICAL EXPOSITIONS have been published.

Volume 1:   MATHEMATICAL GEMS, by Ross Honsberger

Volume 2:   MATHEMATICAL GEMS II, by Ross Honsberger

Volume 3:   MATHEMATICAL MORSELS, by Ross Honsberger

PREFACE

Mathematics abounds in bright ideas. No matter how long and hard one pursues her, mathematics never seems to run out of exciting surprises. And by no means are these gems to be found only in difficult work at an advanced level. All kinds of simple notions are full of ingenuity. The present volume discusses scores of elementary problems which have been culled primarily from the American Mathematical Monthly, 1894–1975. They contain dozens of marvellous ideas, and some twenty or so are just beautiful.

Paul Erdös has the theory that God has a book containing all the theorems of mathematics with their absolutely most beautiful proofs, and when he wants to express particular appreciation of a proof he exclaims, "This is one from the book!" Perhaps it would not be unfitting, at this point, to declare that this book is written from the viewpoint that all the riches of life are gifts from God and that we should receive them with gladness and thanksgiving and share them to His praise and glory.

A knowledge of freshman mathematics more than suffices for most of the book. Occasionally other topics are assumed. Even then it is almost always a standard elementary item which, for lack of space in today's crowded curricula, has had to be placed later in our programs. I refer to such things as Pick's Theorem and the rudiments of circular inversion. There is no cause for alarm if these things are not in your background, for they can readily be picked up when the need arises. References to such items are given in the text.

Most of the problems discussed here have appeared in the problems sections of well-known mathematics journals. At the beginning of each problem, reference is made to its Source, Proposer, and Solver. Accurate credits are often obscured or lost as a problem gets around. Consequently, there is some risk involved in assuming that a problem actually originates with the Proposer and in supposing that the given solution belongs exclusively to the Solver. The references are included mainly to indicate where I happened upon the topic. In fairness to everyone concerned, it should be noted that I have often taken only a part of a problem or solution and I have generally rewritten and embellished all presentations very substantially. The statements of many problems have been revised in the hope of giving them more striking form. This is not a book of problems which are posed for you to solve (although you will undoubtedly have more fun if you try them a little first), but as a showcase for some of mathematics' minor miracles. However, a couple of dozen exercises have been included at the end.

In order to help the reader to locate a particular problem or to pursue a topic of interest, the book concludes with a list of the problems under the three major headings

    (i) Algebra, Arithmetic, Number Theory, Sequences, Probability

    (ii) Combinatorics, Combinatorial Geometry (Maxima and Minima)

    (iii) Geometry (Maxima and Minima).

In the references, the following abbreviations have been used:

AMM—American Mathematical Monthly;

MM—Mathematics Magazine;

NMM—National Mathematics Magazine (forerunner of MM).

I am very grateful to Professor Ivan Niven for a careful review of the manuscript which led to many corrections and other improvements in the final revision. I would also like to thank my colleague Leroy Dickey, and Professors E. F. Beckenbach, Henry Alder, Ralph Boas, Donald Albers, and G. L. Alexanderson for their constructive criticism.

ROSS HONSBERGER

CONTENTS

THE CHESS TOURNAMENT*

There are more chess masters in New York City than in the rest of the U.S. combined. A chess tournament is planned to which all American masters are expected to come. It is agreed that the tournament should be held at the site which minimizes the total intercity traveling done by the contestants. The New York masters claim that, by this criterion, the site chosen should be their city. The West Coast masters argue that a city at or near the center of gravity of the players would be better. Where should the tournament be held?

Solution:

The New Yorkers are right! Suppose the New York masters are denoted $N_1, N_2, \ldots, N_k$, and the other players, in any order, $O_1, O_2, \ldots, O_t$. Since New York has more than half the players, we have $k > t$. In pairing them up $(N_1, O_1), (N_2, O_2), \ldots, (N_t, O_t)$, the New York masters $N_{t+1}, N_{t+2}, \ldots, N_k$ are not involved.

Consider now the pair (N_1, O_1). Wherever the tournament is held the masters N_1 and O_1, between them, must travel a total distance of at least "$N_1 O_1$", the straight-line distance between their cities. Altogether, then, the total distance travelled by all the players is at least

$$S = N_1 O_1 + N_2 O_2 + \cdots + N_t O_t.$$

If the site is New York, S gives the exact amount. However, if the

*Pi Mu Epsilon, Vol. 1, 1949–54, p. 328, Problem 41, proposed and solved by Chester McMaster, New York City.

tournament is held anywhere else, the t pairs of players must travel at least a distance S and the no-longer-zero travel of $N_{t+1}, N_{t+2}, \ldots, N_k$ increases the total. Hence New York is the best site.

A similar problem is considered by J. H. Butchart and Leo Moser in their superb article *No Calculus Please*, Scripta Mathematica, 1952, pp. 221–236:

n points, $x_1, x_2, \ldots, x_n$, are given in order along a line; find the point x on the line at which the sum S of the distances from the given points is a minimum (FIG. 1).

Clearly, the distances $x_1 x$ and $x_n x$ must add up to at least $x_1 x_n$. Now the points go together in pairs "from the outside in" to form a set of nested intervals $(x_1, x_n), (x_2, x_{n-1})$, and so on. If n is odd, the point $x_{(n+1)/2}$ remains unpaired. Since the sum of the distances of the points of a pair is minimized by any point x in between them, a point x in the innermost interval simultaneously minimizes all the pairs of distances. Thus, for n even, we have

$$S \geqslant x_1 x_n + x_2 x_{n-1} + \cdots,$$

with equality for x anywhere in the innermost interval. If n is odd, the same minimum is achieved by taking x at the given point $x_{(n+1)/2}$ (which occurs in the innermost interval), making the extra distance in this case, namely $x x_{(n+1)/2}$, equal to zero.

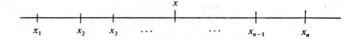

FIG. 1.

THE ORDERED PARTITIONS OF n*

The number 3 can be expressed as the sum of one or more natural numbers in 4 ways, taking into account the order of the terms:

$$3, \quad 1+2, \quad 2+1, \quad 1+1+1.$$

How many such expressions are there for the number n?

Solution:

Consider a string of n 1's in a row. Any arrangement of $n-1$ or fewer dividers inserted in the $n-1$ interior spaces between the 1's corresponds to an expression for n, and conversely.

$$1 \; 1 \mid 1 \; 1 \; 1 \mid \; 1 \mid 1 \; 1 \; \dots \dots 1 \; 1$$
$$n = \; 2 \; + \; 3 \; + 1 + \quad n-6$$

Since we have the choice of inserting or not inserting a divider into each of the $n-1$ spaces, there are 2^{n-1} ways to place the dividers and the same number of expressions for n.

*Pi Mu Epsilon, Vol. 1, 1949–54, p. 186, Problem 27, proposed by Arthur B. Brown, Queen's College, solved by William Moser, University of Toronto.

REGIONS IN A CIRCLE*

Place n points around a circle and draw the chords which join them in pairs. Suppose that no three of the chords are concurrent inside the circle. Into how many regions is the interior of the circle divided?

Solution:

Suppose the chords are added to the figure one at a time. A new chord slices through various regions, increasing the number of sections (FIG. 2). The number of extra regions is the number of segments into which the new chord is divided by the chords it crosses. It is therefore one more than the number of points of intersection which occur along the new chord. From this observation we can easily prove the remarkable formula that the number of regions determined by L lines, no three concurrent inside the circle, which produce P points of intersection in the circle is

$$P + L + 1.$$

For $L = 1$, we have $P + L + 1 = 0 + 1 + 1 = 2$ (FIG. 3). An additional line which crosses the first gives

$$P + L + 1 = 1 + 2 + 1 = 4,$$

and one which does not cross the first yields

$$P + L + 1 = 0 + 2 + 1 = 3.$$

Suppose that $P + L + 1$ is valid for L lines, $L \geqslant 2$. Let an additional line provide k new points of intersection. In crossing k other

*AMM, 1973, p. 561, E2359, proposed by T. C. Brown, Simon Fraser University, Burnaby, British Columbia, solved by Norman Bauman, Nanuet, New York.

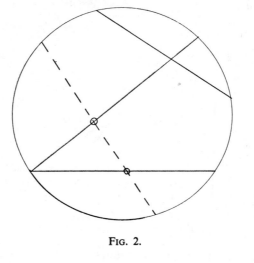

FIG. 2.

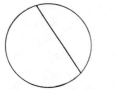

FIG. 3.

lines it goes through $k+1$ regions, increasing the count by $k+1$. Thus, for the $L+1$ lines and $P+k$ points the total number of regions is

$$(P+L+1)+(k+1)=(P+k)+(L+1)+1,$$

establishing the formula by induction.

Clearly there is a 1-1 correspondence between the points of intersection X and the quadruples (A,B,C,D) of the n given points around the circle (FIG. 4). Therefore the number of points of intersection is $\binom{n}{4}$, the number of quadruples. Since there are $\binom{n}{2}$ chords, the number of regions produced by joining the n

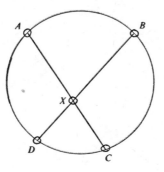

Fig. 4.

points in pairs is

$$P+L+1=\binom{n}{4}+\binom{n}{2}+1.$$

(We observe that our result applies to any convex region of the plane and that the approach generalizes to higher dimensions. A set S of points is *convex* if, for every pair of points A and B in S, the entire segment AB belongs to S. For a detailed account of convex sets, see Russell V. Benson: Euclidean Geometry and Convexity, McGraw-Hill, 1966.)

Observe also that the number of regions can be derived very nicely by keeping track of the number of regions which are lost when the lines are deleted one at a time. Each section of a line separates two regions which unite into a single part when the line is removed. The number of regions lost, then, is one more than the number of points of intersection on the line. Now each point of intersection lies on two lines, and upon the deletion of either of them it vanishes from the other as well. Therefore each point of intersection figures exactly once in the total dismantling operation, and for each line the number of regions lost is

(the number of points of intersection *left* on the line) $+1$.

Summing over the deletion of all L lines, we see that our total will contain all P points of intersection plus a 1 for each line, implying that a total of $P+L$ regions are lost. Since the interior still remains at the end of all this, there must have been $P+L+1$ regions in the first place.

THE FERRY BOATS*

Two ferry boats ply back and forth across a river with constant speeds, turning at the banks without loss of time. They leave opposite shores at the same instant, meet for the first time 700 feet from one shore, continue on their way to the banks, return and meet for the second time 400 feet from the opposite shore. As an *oral* exercise, determine the width of the river.

Solution:

By the time of their first meeting, the total distance that the two boats have traveled is just the width of the river (FIG. 5). It may take one mildly by surprise, however, to realize that, by the time they meet again, the total distance they have traveled is *three* times the width of the river. Since the speeds are constant, the second meeting occurs after a total time that is three times as long as the time for the first meeting. In getting to the first meeting, ferry A (say) traveled 700 feet. In three times as long, it would go 2100 feet. But, in making the second meeting, A goes all the way across the river and then back 400 feet. Thus the river must be 2100 − 400 = 1700 feet wide.

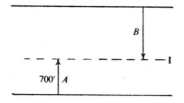
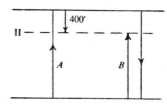

FIG. 5.

*AMM, 1940, p. 111, Problem E366, proposed by C. O. Oakley, Haverford College, solved by W. C. Rufus, Observatory, University of Michigan.

THE BULGING SEMICIRCLE*

A semicircle is drawn outwardly on chord AB of the circle with center O and unit radius. Clearly, the point C on this semicircle which sticks out of the given circle the farthest is on the radius ODC which is perpendicular to AB (FIG. 6). (For any other point C' on the semicircle, we have $OC' < OD + DC' = OD + DC = OC$.) Of course, the size of OC depends on the choice of the chord AB. Determine AB so that OC has maximum length.

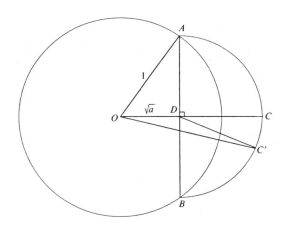

FIG. 6.

*Pi Mu Epsilon, Vol. 4, 1964, p. 355, Problem 187, proposed by R. C. Gebhardt, Parsippany, New Jersey, solved by Murray Klamkin, Ford Scientific Laboratory.

Solution:

Let $OD = \sqrt{a}$. Then the radius of the semicircle is

$$AD = \sqrt{1-a} = DC.$$

Thus $OC^2 = (OD + DC)^2 = (\sqrt{a} + \sqrt{1-a})^2 = a + 2\sqrt{a(1-a)}$ $+ 1 - a$, giving $OC^2 = 1 + 2\sqrt{a(1-a)}$. For a maximum, the value of $a(1-a)$ must be a maximum. Since

$$a(1-a) = a - a^2 = \tfrac{1}{4} - \left(a - \tfrac{1}{2}\right)^2,$$

this occurs for $a = \tfrac{1}{2}$, making $OD = \sqrt{a} = \sqrt{2}/2$.

We observe that, for maximum OC,

$$AD = \sqrt{1 - OD^2} = \sqrt{1 - \tfrac{1}{2}} = \frac{\sqrt{2}}{2},$$

making $AB = 2AD = \sqrt{2}$. Thus $\triangle AOB$ has sides 1, 1, and $\sqrt{2}$, implying that AB subtends a right angle at the center O.

The following clever approach provides an alternative solution. The triangle ADC is right-angled and isosceles, making $\angle DCA = 45°$ (FIG. 7). Now, unless CA is tangent to the given circle, there is a chord which locates the point C farther along the line OD. Thus the chord which maximizes OC must make CA a tangent, in which case CA is a leg of the right-angled isosceles triangle OAC. This makes $CA = OA = 1$ and, from right-angled isosceles triangle DAC, we obtain

$$AD = \frac{\sqrt{2}}{2}, \quad \text{and} \quad AB = \sqrt{2} .$$

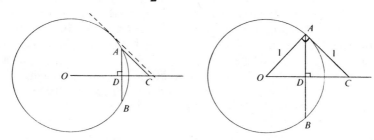

FIG. 7.

THE CHAUFFEUR PROBLEM*

Mr. Smith, a commuter, is picked up each day at the train station at exactly 5 o'clock. One day he arrived unannounced on the 4 o'clock train and began to walk home. Eventually he met the chauffeur driving to the station to get him. The chauffeur drove him the rest of the way home, getting him there 20 minutes earlier than usual.

On another day, Mr. Smith arrived unexpectedly on the 4:30 train, and again began walking home. Again he met the chauffeur and rode the rest of the way with him. How much ahead of usual were they this time? (Assume constant speeds of walking and driving and that no time is lost in turning the car around and picking up Mr. Smith.)

Solution I:

The usual method of solving this problem is the following:

On the first day, the chauffeur was spared a 20-minute drive. Thus Mr. Smith must have been picked up at a point which is a 10-minute drive (one way) from the station. Had the chauffeur proceeded as usual, he would have arrived at the station at exactly 5 o'clock. The 10-minute saving means that he must have picked up Smith at 4:50. Thus Smith took 50 minutes to walk what the chauffeur would take 10 minutes to drive. From this we see that the chauffeur goes 5 times as quickly as Smith.

*This is an old problem with a new twist. In 1974, it appeared on the University of Waterloo's Freshman Mathematics Contest, which was conducted by Murray Klamkin.

Now, on the second day, suppose that Smith walks for $5t$ minutes. The distance he covers, then, would take the chauffeur only t minutes to drive. Accordingly, Smith was picked up this time at t minutes before 5 o'clock, that is, at $60-t$ minutes after 4 o'clock. However, starting at 4:30 and walking for $5t$ minutes, Smith must have been picked up at $30+5t$ minutes after 4 o'clock. Hence $30+5t=60-t$, and $t=5$. Therefore the chauffeur was spared a 5-minute drive (each way), providing a saving of 10 minutes this time.

This is a very nice solution to the problem. However, one Richard Cameron (Peterborough, Ontario), a contestant in our freshman competition, worked out the following solution on the spot during the contest.

Solution II:

Suppose a graph is made by plotting the "distance from the station" against "time." In this way one can easily trace the movements of Smith and the chauffeur. For example, on a normal day we have the following situation, starting from 4 o'clock, say (FIG. 8). (This is the so-called "world line diagram", but Cameron had never heard of such a thing.)

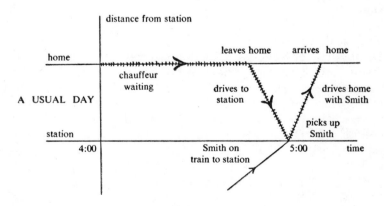

FIG. 8.

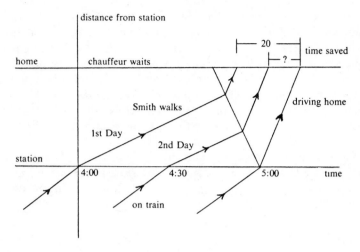

FIG. 9.

The three trips, including a normal day, graphed together give the following picture (FIG. 9). From the constant rates of walking and driving, these world lines are sectionally parallel. Thus, since 4:30 bisects the time between 4 and 5 o'clock, the 1:1 ratio of parts is carried by the parallel lines to give the time saved on the second day to be $\frac{1}{2}(20) = 10$ minutes.

THE SCREENS IN THE CORNER*

Across one corner of a rectangular room are placed two 4-foot screens in such a manner as to enclose the maximum floor space. Determine their positions.

Solution:

The solution is based on the repeated application of the following well-known result:

LEMMA. *The triangle of greatest area in a family of triangles which have a fixed base b and the same angle θ opposite the base is the one which is isosceles. (Clearly, the family fits into a segment of a circle, and the isosceles triangle possesses the greatest altitude to their common base (FIG. 10).)*

Let O denote the corner of the room and suppose the screens are placed to touch the walls at A and B. Then the floor area enclosed is $\triangle AOB + \triangle ABP$, where P denotes the point where the screens meet. (Clearly, the screens must abut for maximum enclosure, making P a common endpoint.)

Now, if $\triangle OAB$ is not isosceles with $OA = OB$, the screens can be moved to positions $A'P'$ and $B'P'$, where $OA' = OB'$ and $A'B'$ remains equal to AB. Then triangles $A'B'P'$ and ABP are equal (in fact, congruent) and, by the lemma, $\triangle OA'B'$ exceeds $\triangle OAB$. The new position, then, encloses a greater area. Thus, in the maximum arrangement, we must have $OA = OB$.

*Pi Mu Epsilon, 1944–45, p. 321, Problem 577, proposed and solved by E. P. Starke, Rutgers University.

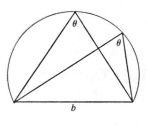
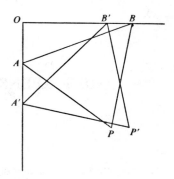

FIG. 10.

Since the screens are of equal length, the point P must always lie on the perpendicular bisector of AB. When $OA = OB$, the perpendicular bisector of AB bisects the right angle in the corner O. Thus, for the greatest area, OP must bisect $\angle O$ (FIG. 11). Hence each screen must maximize the triangle that it cuts from the 45° angle that it subtends at O (lest a better arrangement exist). Since AP is constant, we see, by the lemma, that $\triangle OAP$ is a maximum when it is isosceles and has $OA = OP$. In this case, $\angle OAP = 67\frac{1}{2}$ degrees, and the solution is evident. The simple problem of constructing AP with straightedge and compass is left as an exercise.

A very neat solution to this problem is also given in the later section *No Calculus Please*, page 56.

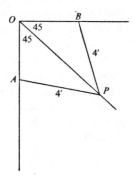
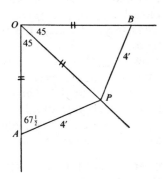

FIG. 11.

COLORING THE PLANE*

Suppose each point of the plane is colored red or blue. Show that some rectangle has its vertices all the same color.

Solution:

Any set of 7 points must contain at least 4 which are the same color. Of 7 points on a line, then, we must have 4 collinear points P_1, P_2, P_3, P_4 which are all the same color, say red. If these points are projected onto two other lines parallel to the first, two collinear quadruples of points $(Q_1, Q_2, Q_3, Q_4), (R_1, R_2, R_3, R_4)$ are obtain-

$$P_1 \cdot \qquad Q_1 \cdot \qquad R_1 \cdot$$
$$P_2 \cdot \qquad Q_2 \cdot \qquad R_2 \cdot$$
$$P_3 \cdot \qquad Q_3 \cdot \qquad R_3 \cdot$$
$$P_4 \cdot \qquad Q_4 \cdot \qquad R_4 \cdot$$

ed which determine several rectangles among themselves and others with the P_i. Now, if any 2 of the Q's are red, an all red rectangle $P_i P_j Q_j Q_i$ results. Similarly for 2 red R's. If neither of these cases hold, then some 3 (or more) of the Q's and some 3 (or more) of the R's must be blue. But these trios of blue points cannot avoid being lined up so that a pair of each trio face each other to yield an all blue rectangle. The conclusion follows.

Observe that the result is valid for any region of the plane which contains all the points inside some circle, no matter how small. Essentially this question occurred on the 5th U.S. Mathematics Olympiad, Spring 1976.

*Pi Mu Epsilon, Vol. 3, 1959–64, p. 474, Problem 138, proposed by David Silverman, Beverly Hills, California, solved by John E. Ferguson, Oregon State University.

AN OBVIOUS MAXIMUM*

P is a variable point on the arc of a circle cut off by the chord AB. Prove the intuitively obvious property that the sum of the chords AP and PB is a maximum when P is at the midpoint of the arc AB.

Solution:

With center O, the midpoint of arc AB, draw a second arc through A and B. Let AP and AO meet this arc, respectively, at Q and C (FIG. 12).

Now AB subtends at the center O an angle which is double the angle it subtends at C on the circumference. But AB subtends the same angle at P as it does at O and the same angle at Q as it does at C. Thus

$$\angle APB = 2\angle Q.$$

But, for $\triangle PQB$, exterior angle APB is equal to the sum of the interior angles at Q and B. Therefore

$$2\angle Q = \angle Q + \angle QBP,$$

and

$$\angle Q = \angle QBP.$$

Thus $\triangle PQB$ is isosceles and $AP + PB = AP + PQ = AQ$, a chord of the outer arc. Clearly this chord is a maximum when it is a diameter, that is, when P is at O.

---

*Pi Mu Epsilon, Vol. 3, 1959–64, p. 296, Problem 130, proposed by H. Kaye, Brooklyn, New York, solved by C. M. Ingleby in the 1860's.

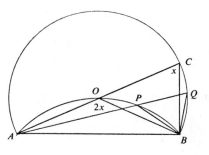

Fig. 12.

Observe that our result also follows from the dynamic consideration of the family of continuously expanding ellipses which have foci at *A* and *B*. By symmetry, the last contact of the ellipses and the circular arc occurs at the midpoint of the arc. The ellipse making this last contact is the largest one to meet the arc and has the greatest sum of focal radii, yielding the desired result.

$$\cos 17x = f(\cos x)^*$$

If f denotes the function which gives $\cos 17x$ in terms of $\cos x$, that is,

$$\cos 17x = f(\cos x),$$

then show that it is the same function f which gives $\sin 17x$ in terms of $\sin x$:

$$\sin 17x = f(\sin x).$$

Solution:

Let $x = \pi/2 - y$. Then $\sin x = \cos y$ and

$$
\begin{aligned}
\sin 17x &= \sin\left[17\left(\frac{\pi}{2} - y\right)\right] \\
&= \sin\left(8\pi + \frac{\pi}{2} - 17y\right) \\
&= \sin\left(\frac{\pi}{2} - 17y\right) \\
&= \cos 17y \\
&= f(\cos y) \\
&= f(\sin x).
\end{aligned}
$$

We observe that the 17 here may be replaced with any integer of the form $4k + 1$.

*MM, 1954, Problem Q103, submitted by Norman Anning.

A SQUARE IN A LATTICE*

An $n \times n$ square S in a coordinate plane covers $(n+1)^2$ lattice points (that is, points whose coordinates (x,y) are both integers) when it is placed so that each corner is on a lattice point and its sides are parallel to the lattice lines (the axes). Prove the highly intuitive result that, no matter how S may be tossed onto the plane, it can never cover more than $(n+1)^2$ lattice points.

Solution:

Consider S in arbitrary position in the plane. Now suppose that the boundary of S is a rubber band and that there is a nail at each lattice point of the plane. Let the rubber band contract, if it can, to fit around the outside of the nails at the lattice points which are covered by S. The polygon H determined by the rubber band is called the "convex hull" of the lattice points of S, and is a concept of greatest importance in many geometric investigations (FIG. 13). If a lattice point occurs at each vertex of S, then H is just S, itself. (See the note at the end of this section for additional comments on the definition of convex hull.)

Because H is contained in S, its area cannot exceed that of S:

$$\text{area of } H \leqslant n^2.$$

Now, in 1899, George Pick discovered a remarkable formula for the area of a polygon Y whose vertices are lattice points and which

---

*AMM, 1968, p. 545, Problem E1954, proposed and solved by D. J. Newman, Yeshiva University.

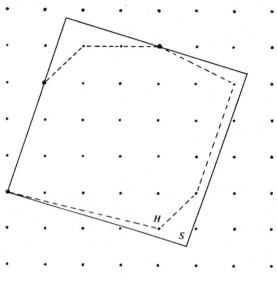

does not cross itself:

$$\text{area of } Y = q + \frac{p}{2} - 1,$$

where q denotes the number of lattice points inside Y and p denotes the number of lattice points on the boundary of Y. (This includes the vertices and any other lattice points on the sides. See the reference for a proof of this theorem.) By Pick's theorem, then, we obtain

$$\text{area of } H = q + \frac{p}{2} - 1 \leqslant n^2,$$

and

$$q + \frac{p}{2} \leqslant n^2 + 1.$$

Because the elastic boundary of S contracted to form H, if it moved at all, the perimeter of H cannot exceed that of S:

$$\text{perimeter of } H \leqslant 4n.$$

Clearly, no two lattice points are closer together than 1 unit. Thus, around the boundary of H there isn't room for a total of more than $4n$ lattice points, giving

$$p \leqslant 4n \qquad \text{and} \qquad \frac{p}{2} \leqslant 2n.$$

Combining this with the earlier result $q + p/2 \leqslant n^2 + 1$, we get

the number of lattice points covered by S

$$= q + p \leqslant n^2 + 1 + 2n = (n+1)^2.$$

Note: The convex hull H of a plane set of points S is the intersection of all the plane convex sets which contain S. H is therefore minimal in the sense that if a convex set contains S it also contains H. H has no superfluous points—it extends only so far as to contain S and to be convex. If S is itself convex, of course H and S are identical.

Reference

Ross Honsberger, Ingenuity in Mathematics, vol. 23, New Mathematical Library, Math. Assoc. of America, 27–31.

AN OPAQUE SQUARE*

A collection of line segments inside or on the boundary of a square of side 1 is said to be "opaque" if every straight line which crosses the square makes contact with at least one of the segments. For example, the two diagonals constitute an opaque set (FIG. 14). Another opaque set is given in (b). The total length of the diagonals is $2\sqrt{2} = 2.82$ approximately, and it is a nice exercise in elementary calculus to prove that the opaque set of minimum total length which has the symmetric pattern illustrated in (b) has length $1 + \sqrt{3} = 2.73$ approximately. Find an opaque set whose length is even less than $1 + \sqrt{3}$.

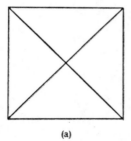 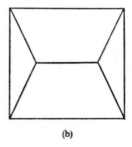

(a)   (b)

FIG. 14.

*AMM, 1966, Mathematical Notes, p. 405, Problem 5.

Solution:

The opaque set consisting of two adjacent sides and the remote half of the diagonal between them (FIG. 15) has length

$$2 + \frac{\sqrt{2}}{2} < 2 + \frac{1.42}{2} = 2.71 < 1 + \sqrt{3} \ .$$

The edges AB and BC make contact with any line that touches $\triangle ABC$ and, with $\triangle ABC$ completely taken care of, the semi-diagonal OD suffices to make the other half of the square opaque.

Now, there is a much more efficient way of looking after $\triangle ABC$. The point P inside $\triangle ABC$ at which each side subtends an angle of 120° is called the Fermat point of the triangle and is the point of the triangle which minimizes the sum of the distances to the vertices (FIG. 16):

$$XA + XB + XC \text{ is a minimum for } X \equiv P.$$

For a thorough treatment of this famous result, see my book *Mathematical Gems*, Vol. 1, Dolciani Mathematical Expositions, Mathematical Association of America, 24–34. It is proved there that the minimum sum $PA + PB + PC$ is given by the segment

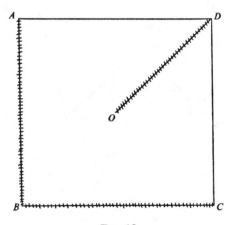

FIG. 15.

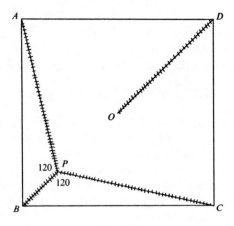

FIG. 16.

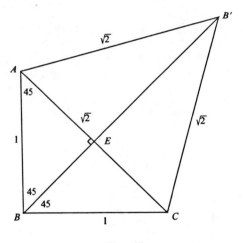

FIG. 17.

BB', where B' is the vertex of the equilateral triangle constructed outwardly on the side AC (FIG. 17). Clearly, in the case at hand, BB' is the perpendicular bisector of AC and $\triangle ABE$ is isosceles. Thus

$$BE = AE = \frac{1}{2}AC = \frac{\sqrt{2}}{2}$$

and

$$BB' = BE + EB' = \frac{\sqrt{2}}{2} + \sqrt{3}\left(\frac{\sqrt{2}}{2}\right) = \frac{\sqrt{2}}{2}(1+\sqrt{3}).$$

Including the semidiagonal $OD = \sqrt{2}/2$, we obtain an opaque set of total length

$$\frac{\sqrt{2}}{2}(2+\sqrt{3}) = 2.64 \text{ approximately.}$$

This splendid solution was pointed out by Maurice Poirier, a secondary school teacher in Orléans, Ontario.

×'S AND O'S*

Suppose a game of ×'s and O's, "tick-tack-toe," is played on an $8 \times 8 \times 8$ cube in 3-dimensional space. How many lines of "8-in-a-row" are there through the cube by which the game might be won?

Solution:

This is not a difficult problem and it yields to a straightforward count. However, a brilliant solution (one of Leo Moser's many) is to consider a $10 \times 10 \times 10$ cube which encases the given $8 \times 8 \times 8$ cube with a shell of unit thickness. The two-way extension of a winning line in the inner $8 \times 8 \times 8$ cube pierces two of the unit cubes in the shell. And each unit cube in the shell is pierced by only one winning line. Thus each winning line corresponds to a unique pair of unit cubes in the outer shell, and the number of winning lines is simply one-half the number of unit cubes in the shell, namely

$$\frac{10^3 - 8^3}{2} = \frac{1000 - 512}{2} = 244.$$

This approach is perfectly general. The number of winning lines for a cube of edge k in n-dimensional space is

$$\frac{(k+2)^n - k^n}{2}.$$

*AMM, 1948, p. 99, Problem E773, proposed by A. L. Rubinoff, University of Toronto, solved by Leo Moser, University of Manitoba.

A SURPRISING PROPERTY OF
RIGHT-ANGLED TRIANGLES*

Prove that if each leg of a right-angled triangle is rotated about its vertex on the hypotenuse so as to lie along the hypotenuse, then the legs overlap in a segment whose length is the diameter of the circle inscribed in the triangle (FIG. 18).

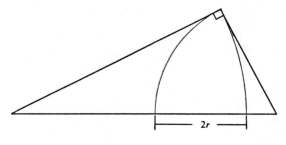

$2r$

FIG. 18.

Solution:

In general, the size of the circle inscribed in a triangle is not a *simple* function of the lengths of the sides. For a right-angled triangle, however, we have the remarkable relation

the diameter = (the sum of the legs) − (the hypotenuse).

---

*AMM, 1956, p. 493, Problem E1197, proposed by Huseyin Demir, Zonguldak, Turkey, solved by Leon Bankoff, Los Angeles, California.

This is easily seen as follows. Radii to the points of contact of the legs project the radius onto each leg (FIG. 19). Since the two tangents to a circle from a point are equal, we have, referring to the diagram,

$$(\text{sum of legs}) - (\text{hypotenuse}) = \left[(x+r) + (r+y) \right] - (x+y) = 2r.$$

The desired conclusion follows the observation that the overlap in question is simply (the sum of the legs) − (the hypotenuse).

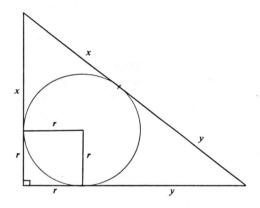

FIG. 19.

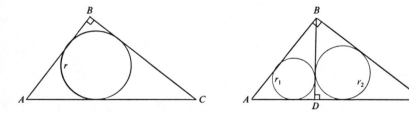

FIG. 20.

The following relation is also easily deduced. If the altitude BD to the hypotenuse of right-angled $\triangle ABC$ is drawn, then the sum of the three in-radii r, r_1, r_2, of $\triangle ABC$ and the smaller triangles into which it is divided by BD, is simply the altitude BD, itself (FIG. 20). We have

$$2r + 2r_1 + 2r_2 = (AB + BC - AC) + (AD + BD - AB)$$
$$+ (BD + DC - BC)$$
$$= (AD + DC) - AC + 2BD = 2BD,$$

hence $r + r_1 + r_2 = BD$.

THE DIGITS OF 4444^{4444}*

The sum of the digits in the decimal representation of 4444^{4444} is A. The sum of the digits in A is B. What is the sum of the digits in B?

Solution:

If the common logarithm of the natural number n lies between $k-1$ and k, then n has k digits in it. Now,

$$\log_{10} 4444^{4444} = 4444 \log_{10} 4444.$$

Using $[x]$ to denote the greatest integer $\leqslant x$, we see, then, that the number of digits in 4444^{4444} is

$$N = [4444 \log_{10} 4444] + 1.$$

Since $4444 < 10^4$, we have $\log_{10} 4444 < 4$ and

$$N \leqslant 4444(4) + 1 < 20{,}000.$$

Because each digit is 9 or less, we have $A < 20{,}000(9) < 199{,}999$. Now the sum of the digits of 199,999 is greater than the sum of the digits of any natural number $< 199{,}999$. Therefore,

$$B = (\text{the sum of the digits in } A) < 1 + 5(9) = 46.$$

The sum S of the digits in B, then, cannot exceed 12, which is the greatest sum of digits obtained by numbers $\leqslant 45$ (given by 39).

Since a number and the sum of its digits are congruent modulo

*17th International Olympiad, 1975, Problem 1.

30

9, we have

$$4444^{4444} \equiv A \equiv B \equiv S \quad (\mathrm{mod}\, 9).$$

Now, modulo 9,

$$4444^{4444} \equiv (-2)^{4444} \equiv (2)^{4444} \equiv 2(2^3)^{1481}$$
$$\equiv 2(-1)^{1481} \equiv -2 \equiv 7.$$

Thus $S \equiv 7$, and this implies $S = 7$ because 7 is the only number up to 12, inclusive, which is congruent to 7 modulo 9.

$$\sigma(n) + \varphi(n) = n \cdot d(n)^*$$

It has often been noted that the remarkable equation $e^{\pi i} = -1$ connects four of the most important numbers in all of mathematics. In a lesser vein, the equation

$$\sigma(n) + \varphi(n) = n \cdot d(n)$$

relates three of the most prominent functions of elementary number theory:

$\sigma(n)$—the sum of the positive divisors of n,

$d(n)$—the number of positive divisors of n,

$\varphi(n)$—Euler's φ-function, that is, the number of natural numbers $m \leqslant n$ which are relatively prime to n, i.e., for which $(m,n) = 1$.

Of course, one can throw together whatever functions he will in order to concoct a mathematical condition. Prove, however, the surprising fact that the equation $\sigma(n) + \varphi(n) = n \cdot d(n)$ is a necessary and sufficient condition for n to be a prime number.

Solution:

(i) Suppose n is a prime number. Then the divisors of n are 1 and n, and we have $\sigma(n) = n+1$, $\varphi(n) = n-1$, and $d(n) = 2$. In this case

$$\sigma(n) + \varphi(n) = 2n = n \cdot d(n),$$

---

*AMM, 1965, p. 186, Problem E1674, proposed by C. A. Nicol, University of South Carolina; solution by Ivan Niven, University of Oregon (unpublished).

implying the condition is necessary.

(ii) Suppose that $\sigma(n) + \varphi(n) = n \cdot d(n)$ and also that n is not a prime number. The equation is not satisfied by $n = 1$ ($1 + 1 \neq 1 \cdot 1$), implying $n \geq 2$.

For $n > 1$, $\varphi(n)$ does not count the number n, itself, and we have

$$\varphi(n) < n.$$

Because n is a composite number it must have at least 3 divisors. Let $d(n)$ be denoted by k, and let the positive divisors of n be denoted by

$$d_1 = 1 < d_2 < \cdots < d_k = n.$$

Since $k = d(n) \geq 3$, the divisor d_2 is not the greatest divisor, yielding

$$d_2 < n \quad \text{and} \quad n - d_2 \geq 1.$$

Consequently, we have

$$\begin{aligned}
n \cdot d(n) - \sigma(n) &= kn - (d_1 + d_2 + \cdots + d_k) \\
&= (n - d_1) + (n - d_2) + \cdots + (n - d_k) \\
&\geq (n - d_1) + (n - d_2) + (n - d_k) \\
&\geq (n - 1) + 1 + 0 \\
&= n \\
&> \varphi(n),
\end{aligned}$$

showing the impossibility of $n \cdot d(n) - \sigma(n) = \varphi(n)$. This contradiction proves that n must be a prime number.

Probably the most famous primality condition of all is Wilson's theorem:

n divides $(n-1)! + 1$ if and only if n is a prime number.

In 1965, the AMM carried the following problem as E1702, proposed by Douglas Lind, Falls Church, Virginia, and solved by Kenneth Kramer, Columbia College, and Steven Minsker, Brooklyn College:

Prove that n divides $N = \sum_{r=1}^{n-3} r(r!)$ if and only if n is a prime number.

Solution:

We have $N = 1(1!) + 2(2!) + \cdots + (n-3)[(n-3)!]$. Since $r(r!) = (r+1)r! - r! = (r+1)! - r!$, we have

$$N = (2! - 1!) + (3! - 2!) + (4! - 3!) + \cdots + \left[(n-2)! - (n-3)!\right]$$
$$= (n-2)! - 1.$$

Multiplying through by $n-1$, and adding n to each side, we obtain

$$(n-1)N + n = (n-1)! + 1.$$

By Wilson's theorem, n is a prime if and only if n is a divisor of $(n-1)! + 1$, and the last equation reveals that this is true if and only if n is a divisor of N, since n and $n-1$ are always relatively prime.

ON k-CLOUDS*

Circles of unit radius are packed, without overlapping of interior points, in a strip S of the plane whose parallel edges are a distance w apart. We say that the circles form a *k-cloud* if every straight line which cuts across S makes contact with at least k circles. Prove that for a 2-cloud, $w \geqslant 2 + \sqrt{3}$ (FIG. 21).

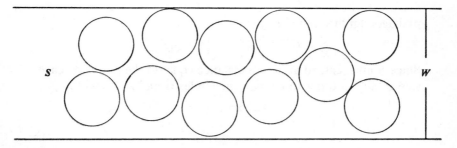

FIG. 21.

Solution:

Through the center O of any circle C in a 2-cloud construct the straight line m which crosses S at right angles (FIG. 22). Now m must make contact with a second circle A. Let Q denote the foot of the perpendicular from the center P of A to the line m. Since m makes contact with A, the length of PQ cannot exceed the radius, that is, $PQ \leqslant 1$. And since C and A do not overlap, $OP \geqslant 2$. The

*AMM, 1966, Mathematical Notes, p. 404, Problem 5.

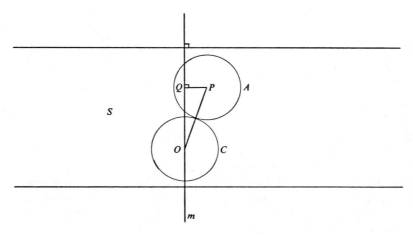

FIG. 22.

pythagorean theorem then yields

$$OQ = \sqrt{OP^2 - PQ^2} \geqslant \sqrt{2^2 - 1^2} = \sqrt{3} .$$

Since S must extend at least a distance equal to the radius on each side of OQ in order to contain the circles C and A, its width

$$w \geqslant 2 + OQ \geqslant 2 + \sqrt{3} .$$

A MINIMAL SUM*

There are n^k different k-tuples $(a_1, a_2, \ldots, a_k)$ which can be constructed by taking a_i from the set $(1, 2, 3, \ldots, n)$, repeated values of the a_i permitted. For each of these k-tuples, the minimum a_i is noted. Prove the surprising result that the sum of all these minimum a_i is simply

$$1^k + 2^k + 3^k + \cdots + n^k,$$

the sum of the kth powers of the first n natural numbers,

$$\text{i.e., } \Sigma \min(a_1, a_2, \ldots, a_k) = \Sigma_{m=1}^n m^k.$$

Solution:

This splendid solution is based on a very primitive notion which is usually too simple to be of much help. In adding up a series of natural numbers r_i, one may distribute his counting by determining the number of terms ≥ 1, the number of terms ≥ 2, and so on.

$$\Sigma r_i = (\text{number of terms } \geq 1) + (\text{number of terms } \geq 2) + \cdots.$$

In this way, a number r is counted in r terms, thus contributing the correct amount to the total. For example, the number $r = 3$ is

---

*AMM, 1976, p. 61, Problem E2507, proposed by R. L. Graham, Bell Telephone Laboratories, Murray Hill, New Jersey, solved by Peter G. de Buda, undergraduate, University of Toronto.

counted in

(number of terms ≥ 1), (number of terms ≥ 2),

(number of terms ≥ 3),

but not in any later term. This procedure is equivalent to passing over the series, picking up 1 from each term (so long as a term has anything left to give), and adding up the totals obtained on the various passes. (It is like considering the term r to be a pile of r matchsticks.)

It turns out to be very easy to calculate the number of terms of the sum in question which are at least equal to the positive integer t. Clearly we have

$$\min(a_1, a_2, \ldots, a_k) \geq t$$

if and only if every a_i in the k-tuple $\geq t$. The number of k-tuples which have this property is simply the number of k-tuples which can be constructed by taking the a_i from the range $(t, t+1, \ldots, n)$. Accordingly, there are $n - t + 1$ choices for each a_i, and the number of such k-tuples is $(n - t + 1)^k$. Therefore

$$\text{(the number of terms} \geq t) = (n - t + 1)^k,$$

and the sum in question is

$$\Sigma \min(a_1, a_2, \ldots, a_k) = \Sigma_{t=1}^n \text{ (the number of terms} \geq t)$$
$$= \Sigma_{t=1}^n (n - t + 1)^k$$
$$= n^k + (n-1)^k + (n-2)^k + \cdots + 1^k.$$

Professor Ivan Niven pointed out the following nice completion of the above argument, carrying on from the point of having established that the number of k-tuples with minimum $\geq t$ is $(n - t + 1)^k$. This formula gives the number of k-tuples with minimum $\geq t + 1$ to be $(n - t)^k$. Therefore the number of k-tuples with minimum exactly t is $(n - t + 1)^k - (n - t)^k$, and these contribute $t[(n - t + 1)^k - (n - t)^k]$ to the total in question. Adding for $t =$

$1, 2, \ldots, n$, the total is seen to be

$$1\left[(n)^k - (n-1)^k\right] + 2\left[(n-1)^k - (n-2)^k\right] + 3\left[(n-2)^k - (n-3)^k\right]$$
$$+ \cdots + n\left[1^k - 0^k\right] = n^k + (n-1)^k + (n-2)^k + \cdots + 1^k.$$

As an exercise, the reader might enjoy determining the value of $\Sigma \max(a_1, a_2, \ldots, a_n)$.

THE LAST THREE DIGITS OF 7^{9999}*

What are the last three digits of 7^{9999}?

Solution:

Observe that $7^4 = 2401$. Therefore

$$7^{4n} = (2401)^n = (1 + 2400)^n = 1 + n \cdot 2400 + \binom{n}{2} \cdot 2400^2 + \cdots,$$

where all the terms in this binomial expansion, beyond the second, end in at least 4 zeros and do not affect the last 3 digits of the result. The last 3 digits are determined by

$$1 + n \cdot 2400 = 24n \cdot 100 + 1.$$

If m is the last digit of $24n$, we have

$$24n \cdot 100 + 1 = (\ldots m)100 + 1 = \ldots m01, \text{ ending in } m01.$$

For $n = 2499$, $24n$ ends in 6, and we see that

$$7^{4n} = 7^{9996} \text{ ends in } 601.$$

Since $7^3 = 343$, we obtain

$$7^{9999} = 7^{9996} \cdot 7^3$$
$$= (\ldots 601)(343) = \ldots 143,$$

giving the last 3 digits to be 143 (by direct multiplication).

Alternatively, we might have proceeded as follows. For $n = 2500$,

---

*NMM, 1937–38, p. 415, Problem 216, proposed by Victor Thébault, Le Mans, France, solved by D. P. Richardson, University of Arkansas.

$24n$ ends in 0 and we see that $7^{4n} = 7^{10000}$ ends in 001. Thus

$$7^{10000} = \ldots 001 = \ldots 000 + 1 = 1000k + 1$$

for some integer k. This yields $7^{10000} = 1000(k-1) + 1001$. Dividing by 7, we get

$$7^{9999} = \frac{1000(k-1)}{7} + 143.$$

Since the right-hand side must be an integer, 7 must divide $1000(k-1)$. But 7 does not divide 1000. Thus it must divide $k-1$ and, for some integer q, we have $7^{9999} = 1000q + 143$. Since $1000q$ ends in 000, we see that 7^{9999} ends in 143.

ROLLING A DIE*

A normal die bearing the numbers 1, 2, 3, 4, 5, 6 on its faces is thrown repeatedly until the running total first exceeds 12. What is the most likely total that will be obtained?

Solution:

Consider the throw before the last one. After this throw the total must be either 12, 11, 10, 9, 8, or 7. If it is 12, then the final result will be either 13, 14, 15, 16, 17, or 18, with an equal chance for each. Similarly, if the next to last total is 11, the final result is either 13, 14, 15, 16, or 17, with an equal chance for each; and so on. The 13 appears as an equal candidate in every case, and is the only number to do so. Thus the most likely total is 13.

In general, the same argument shows the most likely total that first exceeds the number n ($n \geqslant 6$) is $n + 1$.

*AMM, 1948, p. 98, Problem E771, proposed by C. C. Carter, Bluffs, Illinois, solved by N. J. Fine, University of Pennsylvania.

PIERCING A CUBE*

A solid cube C of dimensions $20 \times 20 \times 20$ is built out of 2000 bricks of dimensions $2 \times 2 \times 1$. Prove that the cube can be pierced by a straight line perpendicular to a face, passing through the interior of the cube, which does not pierce any of the bricks.

Solution:

Consider the 8000 unit cubes in C. On each face of C, the edges of the unit cubes outline a 20×20 grid. In the interior of a face there are $19(19) = 361$ points in this grid. The 361 lines through these points, perpendicular to the face, run along the edges of the rows of unit cubes to the corresponding grid points in the opposite face. Altogether there are $3(361) = 1083$ such lines piercing the interior of C. Let L denote an arbitrary member of this set of lines.

The two planes through L, parallel to faces of C, divide C into four rectangular sections, each of which has L as an edge. Let A denote any of these sections (FIG. 23). Since one of the dimensions of A is 20 (a whole edge of C), A must contain an even number of unit cubes. Now, a $2 \times 2 \times 1$ brick may have 1, 2, or 4 of its unit cubes in A, but no brick can contribute exactly 3 unit cubes to A. The total contribution to A of the bricks providing 2 or 4 unit cubes is an even number of unit cubes. Therefore there must be an even number of bricks which give exactly 1 unit cube to A. Now a brick which has exactly 1 unit cube in A must have L run right

*AMM, 1971, p. 801, Problem 5744, proposed by Jan Mycielski, University of Colorado, solved by Bill Sands, University of Manitoba.

43

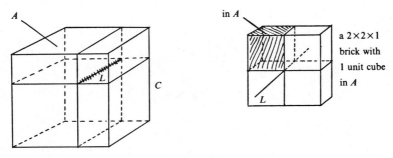

FIG. 23.

through the middle of it, along the edge which is common to its four unit cubes. Consequently, L must pierce an *even* number of bricks.

But a given brick can be pierced by only one line L. In fact, each brick is pierced by one and only one line L, providing a total of 2000 such piercings. Since there are 1083 lines L, each of which pierces an even number of bricks, not all of the lines L can pierce as many as 2 bricks. At least 83 of the lines L must pierce no brick at all.

DOUBLE SEQUENCES*

For certain natural numbers n it is possible to construct a sequence in which each number $1, 2, 3, \ldots, n$ occurs twice, the second occurrence of each number r being r places beyond its first occurrence. For example, for $n = 4$,

$$4, 2, 3, 2, 4, 3, 1, 1.$$

For $n = 5$, we have

$$3, 5, 2, 3, 2, 4, 5, 1, 1, 4.$$

Such arrangements do not exist for $n = 6$ or 7. For $n = 8$ we have

$$8, 6, 4, 2, 7, 2, 4, 6, 8, 3, 5, 7, 3, 1, 1, 5.$$

Prove that such a sequence cannot exist unless $n \equiv 0$ or $1 \pmod 4$.

Solution:

Let the positions in such a sequence be numbered $1, 2, 3, \ldots, 2n$. Suppose that the first occurrence of the number 1 is in position p_1, the first occurrence of 2 is in position p_2, and so on. Then the second occurrence of 1 is in position $p_1 + 1$, and the second appearance of 2 is in position $p_2 + 2$, etc. The sum of the numbers denoting the positions is then both

$$1 + 2 + 3 + \cdots + 2n$$

and $(p_1 + p_1 + 1) + (p_2 + p_2 + 2) + \cdots + (p_n + p_n + n)$. Therefore

$$\frac{2n(2n+1)}{2} = 2(p_1 + p_2 + \cdots + p_n) + \frac{n(n+1)}{2}.$$

*AMM, 1967, p. 591, Problem E1845, proposed by R. S. Nickerson, Hanscom Field, Bedford, Massachusetts, solved by D. C. B. Marsh, Colorado School of Mines.

Letting $p_1 + p_2 + \cdots + p_n = P$, we get

$$\frac{2n(2n+1)}{2} = 2P + \frac{n(n+1)}{2},$$

giving

$$P = \frac{2n(2n+1) - n(n+1)}{4} = \frac{3n^2 + n}{4} = \frac{n(3n+1)}{4}.$$

Since P is an integer, $n(3n+1)$ must be divisible by 4. However, for $n \equiv 2$ or $3 \pmod 4$, we have $n(3n+1) \equiv 2 \pmod 4$, a contradiction. Therefore there can be no sequence unless $n \equiv 0$ or $1 \pmod 4$.

In his solution (see reference), D. C. B. Marsh proves that a sequence does exist for every $n \equiv 0$ or $1 \pmod 4$.

Consider now the companion problem of determining the number of ways of arranging the numbers $1, 2, \ldots, n$ in a row so that, except for an arbitrary choice of first number, the number k can be placed in the row only if it is preceded either by $k-1$ or $k+1$ (not necessarily immediately). For example, for $n = 6$, among the acceptable arrangements are

4 3 5 2 6 1   and   3 4 2 1 5 6.

Solution:

Suppose the first number is r. This splits the numbers $1, 2, \ldots, n$ into two groups

$$A = (1, 2, \ldots, r-1) \quad \text{and} \quad B = (r+1, r+2, \ldots, n).$$

The rule requiring either $k-1$ or $k+1$ ahead of k implies that the numbers in A must occur in the row in their natural descending order and that the numbers in B occur in ascending order. It is clear that the first number from B, for example, which can appear after the initial value r, is not $r+2$ because it would not be preceded by either $r+1$ or $r+3$. Similarly, none of $r+3, r+4, \ldots, n$ can be the first member of B to follow r. Repeated application of this argument shows that the members of B must occur in the order $r+1, r+2, \ldots, n$. Similarly for A.

So long as the elements of A and B keep their natural order within their own group, it doesn't matter how the groups may be interlaced. For example, since r is first, the $r+1$ can go in right away or later, say after the $r-1$ and $r-2$ have been inserted. Consequently, the number of rows which begin with r is the number of ways of choosing $r-1$ places from the remaining $n-1$ places for the $r-1$ members of A, namely $\binom{n-1}{r-1}$. The ordered members of B automatically go in the vacant places. Since r may take the values $1, 2, \ldots, n$, the total number of ways of arranging the row is

$$\binom{n-1}{0} + \binom{n-1}{1} + \binom{n-1}{2} + \cdots + \binom{n-1}{n-1} = (1+1)^{n-1} = 2^{n-1}.$$

POINT-SPLITTING CIRCLES*

$2n + 3$ points are given in the plane, no 3 on a line and no 4 on a circle. Prove that it is always possible to find a circle which goes through 3 of the given points and splits the others in half, that is, has n on the inside and n on the outside.

Solution I: By Murray Klamkin, University of Waterloo.

Let A and B denote two consecutive vertices of the convex hull H of the given set of points T (FIG. 24). (See Problem 11, pp. 19–21 for the definition of convex hull.) Since no three of the points of T lie on a line, there are no points of T on the segment AB between A and B. Let a large circle C', through A and B, have its center O outside H. Now, the segment AXB of C', which lies inside H, may contain some of the points of T. However, by increasing the size of C', the arc AXB can be made to approach AB as closely as desired. In such an expansion, the arc AXB, in approaching AB, would pass over all the (finite number of) points of T which were in the original segment AXB. Thus there exists any number of circles through A and B for which the segment AXB, including its arc, is devoid of points of T. Let C denote such a circle and let its center be O.

To begin with, then, we have a circle C which passes through exactly two points of T, namely A and B, and has no points of T inside it. Let C be transformed by moving its center O toward AB along the perpendicular bisector L of AB, adjusting the radius all

---

*This problem comes from a 1962 mathematics competition held in China; see AMM, 1972, p. 899, The Chinese Mathematical Olympiads, Frank Swetz, Pennsylvania State College.

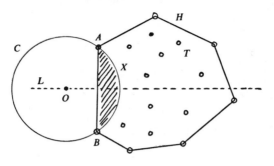

Fɪɢ. 24.

the while so as to keep C passing through A and B. As O
approaches AB, the segment AXB enlarges to take in more and
more of H. When O crosses AB, the segment AXB becomes the
major segment of C (determined by AB) and eventually grows to
cover the entire hull H, and therefore the entire set T. Since no 4
points of T lie on a circle, the segment AXB must assimilate the
points of T *one at a time*. When the $(n+1)$th point K is picked up
by the circumference, the circle C will contain precisely the
preceding n points inside it and will pass through the three points
A, B, and K. The other n points of T are still beyond the confines
of C, and the conclusion follows.

Solution II: By L. J. Dickey, University of Waterloo.

Let O denote any of the given points and consider any circle R
which has center O. Let the given points be subjected to a circular
inversion in R. (See the reference for an account of circular
inversion.) The center O is thus carried to the point at infinity,
yielding a set S of $2n+2$ finite image points (Fɪɢ. 25).

Beginning with a line t which has all of S on one side, move t
toward S until it first makes contact with S, say at a point A'.
(Alternatively: let A' denote a vertex of the convex hull of S and
let t be a line through A' which meets the hull only at A'.) Now let
either ray of t, determined by A', sweep across S by being pivoted
about A'. We will prove shortly that this sweeping ray of t will
meet the points of S *one at a time*. Consequently, it eventually

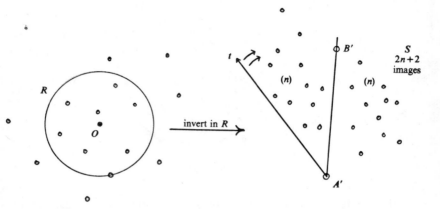

FIG. 25.

reaches an image point B' such that $A'B'$ splits the other $2n$ points of S in half.

Let A and B of the given set denote the antecedents of A' and B'. We observe that $A'B'$ cannot pass through the center of inversion O, for this would mean that A, B, and O of the given set would lie on a line (a contradiction). Therefore the antecedent of the line $A'B'$ must be a circle K through O (FIG. 26). K, then, passes through the three points A, B, and O of the given set. And because, when inverting back the images of S to the points of the given set, all the images on one side of $A'B'$ are taken inside K and all those on the other side of $A'B'$ are carried outside K, we see that K splits the given set in half. (The tangent to K at O is parallel to $A'B'$.)

Not only have we solved the problem, but we see that through each point O of the given set a solution circle can be found. Of $2n+3$ circles thus obtained, there may be only $(2n+3)/3$ different circles, for the one through O, A, and B may be counted 3 times. (Since no 4 points of the given set lie on a circle, no solution circle can be counted more than 3 times in these $2n+3$ solutions.) It is not impossible for a given point to have more than one solution circle pass through it.

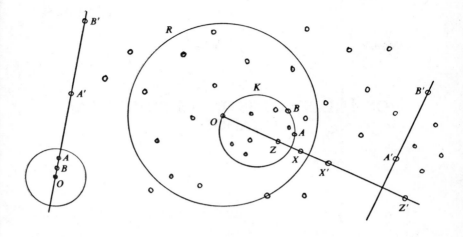

We conclude by justifying the earlier claim that the sweeping ray of t encounters the images in S one at a time. First, we observe, as before, that whenever the ray meets an image point C', it cannot, in this position, also pass through the center O (if it did, then O and the antecedents C and A would constitute three given points on a line). Thus, if the ray were to encounter simultaneously two (or more) images, say C' and D', the antecedent of the line $A'C'D'$ would be a circle through O (since $A'C'D'$ does not go through O), containing the *four* given points $O, A, C,$ and D (a contradiction).

Reference

Coxeter and Greitzer, Geometry Revisited, vol. 19, New Mathematical Library, Math. Assoc. of America, p. 108 ff.

ON THE LENGTHS OF THE SIDES OF A TRIANGLE*

If a, b, c form a triangle, prove that, for all $n = 2, 3, 4, \ldots,$ $\sqrt[n]{a}, \sqrt[n]{b}, \sqrt[n]{c}$ also form a triangle.

Solution:

If a, b, c form a triangle, the triangle inequality yields
$$a + b > c,$$
and so on. Therefore we have
$$\left(\sqrt[n]{a} + \sqrt[n]{b} \right)^n > a + b > c = \left(\sqrt[n]{c} \right)^n,$$
giving
$$\sqrt[n]{a} + \sqrt[n]{b} > \sqrt[n]{c}.$$
The other parts of the triangle inequality for $\sqrt[n]{a}, \sqrt[n]{b}, \sqrt[n]{c}$ are similarly established, and the conclusion follows.

*AMM, 1960, p. 82, Problem E1366, proposed by V. E. Hoggatt, Jr., solved by R. T. Hood, Ohio University.

NO CALCULUS PLEASE

In 1952, J. H. Butchart and Leo Moser published an exceptional article *No Calculus Please* in the popular journal Scripta Mathematica, pages 221–236. We consider here a few of the ingenious alternatives they present to the usual calculus approach.

(i) Our first problem is to determine the volume common to two right circular cylinders of radius "a" which intersect at right angles (that is, their central axes intersect at right angles). It gives one a moment's pause to picture the intersection R just described. In the direction of one of the cylinders, R appears circular, being rounded as the cylinder from which it is cut. The top half of R is sketched below (FIG. 27). Its volume is found very easily by a comparison with its inscribed sphere. Clearly, a sphere of radius a could roll down either cylinder and is the inscribed sphere of their intersection.

Because of its 4-fold symmetry, any slice of R, parallel to the base $ABCD$, would have a square base. Such a slice of the inscribed sphere S would have a circular base which would fit tangentially inside the square base of the corresponding slice of R. The ratio of the area of a square to that of its inscribed circle is

$$\frac{(2r)^2}{\pi r^2} = \frac{4}{\pi}.$$

Summing over all levels, we obtain the volume of R to be $4/\pi$ times that of S. Thus

$$\text{the volume of } R = \frac{4}{\pi}\left(\frac{4}{3}\pi a^3\right) = \frac{16}{3}a^3.$$

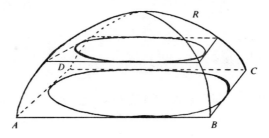

FIG. 27.

Of course, a rigorous justification of $R = 4S/\pi$ would involve the basic calculus arguments this approach purports to avoid. There is really no way to get around the notion of limit in such a problem. Nevertheless, the comparison of R with S is ingenious, satisfying, and can be justified rigorously.

(ii) Another clever idea concerns the problem of determining the line L through a given point P inside a convex curve C which cuts from C a region of minimum area. Unless P bisects L, a slight rotation about P, in the appropriate direction, causes the region to lose a larger wedge on one side of P than it gains on the other side, showing that the region is not minimal (FIG. 28). This is an old idea that applies to many problems.

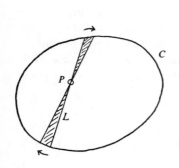

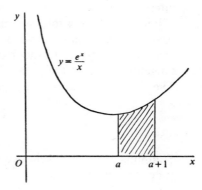

FIG. 28.

An attractive application of this notion provides an immediate solution to the problem of placing two ordinates, one unit apart, to the curve $y = e^x/x$ so that they determine a region under the curve of minimum area. Clearly, if the two ordinates are not equal, a slight shift toward equality gives a better result. For the minimum, then, we must have, for some real number a,

$$\frac{e^a}{a} = \frac{e^{a+1}}{a+1}, \qquad \text{giving } a = \frac{1}{e-1}.$$

(iii) Moser and Butchart cleverly exploit the classic isoperimetric result that of all simple n-gons with given perimeter L, the one of greatest area is regular (simple means it does not cross itself). A neat new proof, which makes the not unusual assumption that such a maximal n-gon exists, was given by the late Richard DeMar (University of Cincinnati) in his outstanding paper *A Simple Approach to Isoperimetric Problems in the Plane*, Mathematics Magazine, 1975, pp. 1–12, in particular pp. 4–6. He considers separately the two requirements for regularity—(i) equal sides, and (ii) equal angles. The cases are handled similarly by a beautiful argument which we shall illustrate by working through part (i).

Let $K = ABCD\ldots$ denote a simple n-gon which is not equilateral (FIG. 29). In this case, some pair of consecutive sides must be unequal. Suppose $AB > BC$. Let X be chosen between A and B such that $BX > BC$. Then the angle $u = \angle BCX$ exceeds the angle $v = \angle BXC$. Now, cut off $\triangle BCX$ and turn it over so as to interchange X and C, producing a polygon K'. Because $u > v$, the angle at X now exceeds a straight angle and K' is nonconvex. Observe also that K' is not an n-gon, but an $(n+1)$-gon. However, because it is nonconvex, we can easily convert it to an n-gon and at the same time improve on its area. By simply joining B', the new position of B, to A, we obtain an n-gon which has a greater area with an even lesser perimeter. Dilating the figure to make the perimeter equal to L increases the area still further, showing that K is not the desired maximum.

Let us assume this isoperimetric theorem and see how Butchart and Moser apply it to the old problem of maximizing the area of a rectangular region which is to be bounded on one side by a fixed

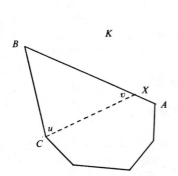
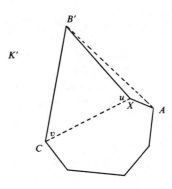

FIG. 29.

wall and on the other three sides by a prescribed total length L of fencing. For all fences $ABCD$, the quadrilateral $B'BCC'$, obtained by reflecting the rectangle in the wall, has constant perimeter $2L$ (FIG. 30). Thus it is maximal when it is regular, namely a square. Accordingly, the maximum "half", $ABCD$, must have its length equal to twice its width, and the problem is solved.

A similar problem requires the tin can of maximum volume, given a prescribed surface area. If the solution is known for either of the cases "with lid" or "without lid," the solution to the other case is quickly determined. For, reflecting an open tin in its upper surface provides a closed tin which must also be maximal. Thus, whatever the ratio h/r of height to radius might be, we see that it is twice as great in the case of the tin "with lid" as it is for the tin "without lid."

(iv) As a final application of this isoperimetric result, let us derive another solution to our Problem 7, page 13:

Across one corner of a rectangular room are placed two 4-foot screens in such a manner as to enclose the maximum floor space. Determine their positions.

Let the corner and walls of the room be taken as the origin and axes of a rectangular frame of reference. Let the screens be reflected in the axes to yield the octagon $PBP_1A_1P_2B_1P_3A$, as shown (FIG. 31). For all positions of the screens, the octagon has

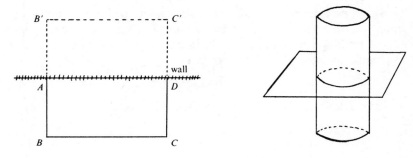

FIG. 30.

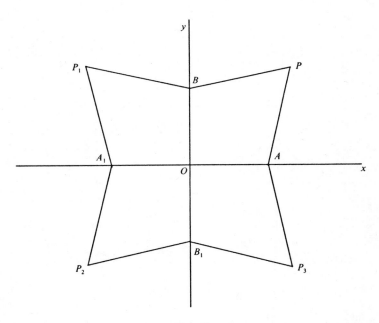

FIG. 31.

constant perimeter $8 \cdot 4 = 32$ feet. The solution follows quickly upon the observation that the maximum floor space in the corner corresponds to the octagon of maximum area. Since this must be the regular octagon, each of its angles is

$$\tfrac{1}{8}(8-2) \cdot 180 = 135 \text{ degrees,}$$

and the details of construction are easily deduced.

a^b AND $b^{a}*$

There is often no difficulty in determining which of a^b and b^a is greater. It is obvious that

$$2^3 < 3^2 \quad \text{and} \quad 3^4 > 4^3.$$

However, coupling a number between 2 and 3 with one between 3 and 4 presents a problem. Which is greater

$$e^\pi \quad \text{or} \quad \pi^e?$$

Solution:

For positive x, we have

$$e^x = 1 + x + \frac{x^2}{2!} + \cdots$$
$$> 1 + x.$$

Since $\pi > e$, we have $\pi/e > 1$, and $x = \pi/e - 1 > 0$. Thus

$$e^{(\pi/e)-1} > 1 + \left(\frac{\pi}{e} - 1\right)$$
$$\frac{e^{(\pi/e)}}{e} > \frac{\pi}{e}$$
$$e^{(\pi/e)} > \pi$$
$$e^\pi > \pi^e.$$

It is an easy exercise in calculus to show that the function $e^x - 1 - x$ has a single minimum value of zero (at $x = 0$). Thus for

*Two-Year College Mathematics Journal, Vol. 6, May 1975, p. 45, "Two More Proofs of a Familiar Inequality", by Erwin Just and Norman Schaumberger, Bronx Community College.

all real values of x we have

$$e^x \geqslant 1 + x.$$

With this relation, George Polya constructed the following beautiful proof of the famous arithmetic-geometric mean inequality. Let $a_1, a_2, \ldots, a_n$ denote positive real numbers and let A and G, respectively, denote their arithmetic and geometric means, that is,

$$A = \frac{1}{n}(a_1 + a_2 + \cdots + a_n), \qquad G = (a_1 a_2 \cdots a_n)^{1/n}.$$

Giving x in turn the values $(a_i/A) - 1$, $i = 1, 2, \ldots, n$, we obtain the n relations

$$e^{(a_1/A) - 1} \geqslant \frac{a_1}{A},$$

$$e^{(a_2/A) - 1} \geqslant \frac{a_2}{A},$$

$$\cdots \cdots \cdots$$

$$e^{(a_n/A) - 1} \geqslant \frac{a_n}{A}.$$

Multiplying all these together we get

$$e^{\frac{a_1 + a_2 + \cdots + a_n}{A} - n} \geqslant \frac{a_1 a_2 \cdots a_n}{A^n},$$

which is simply

$$e^{n-n} \geqslant \frac{G^n}{A^n}, \quad \text{or} \quad 1 \geqslant \frac{G^n}{A^n},$$

from which it follows that $A \geqslant G$.

Observe that $A = G$ only if equality holds in all n relations. This requires $a_i/A - 1 = 0$ in all cases, showing that $A = G$ only when all the a_i are equal (to A).

A MATHEMATICAL JOKE*

A man purchased at a post office some one-cent stamps, three-quarters as many two's as one's, three-quarters as many five's as two's, and five eight-cent stamps. He paid for them with a single bill, and there was no change. How many stamps of each kind did he buy?

Solution:

Suppose the man bought y one-cent stamps. Then he bought $3y/4$ two-cent stamps and $9y/16$ five-cent stamps. Since $9y/16$ is an integer, 16 must divide y. For some integer x, then, we have $y = 16x$ and the purchase consisted of $16x$ 1's, $12x$ 2's, $9x$ 5's, and 5 8's. Suppose he paid for the stamps with a k-dollar bill. Then the cost of the stamps is

$$16x + 2(12x) + 5(9x) + 8(5) = 100k,$$

giving

$$85x + 40 = 100k,$$
$$17x = 20k - 8,$$

and

$$x = \frac{20k - 8}{17} = k + \frac{3k - 8}{17}.$$

---

*AMM, 1936, p. 48, Problem E163, proposed by W. A. Carver, Lakewood, Ohio, solved by C. C. Richtmeyer, Mt. Pleasant, Michigan.

Since x and k are integers, it follows that

$$\frac{3k-8}{17} \text{ must also be a whole number.}$$

But k must be one of the numbers 1, 2, 5, 10, 20, 50, 100, 1000, or 10000. The only value of k which makes $(3k-8)/17$ a whole number is $k = 1000$, implying that the man bought

18,816 one's, 14,112 two's, 10,584 five's, and 5 eight's.

MAPS ON A SPHERE*

Consider a map M on a sphere. Suppose that there are exactly 3 countries at each vertex of M and that none of its boundaries are loops (that is, each boundary contains at least 2 vertices). The map is colored with colors A, B, C, D so that different colors are used for countries which share a length of common border. We use the term "country" for all the regions in the map; we don't care which are really countries and which are bodies of water.

A country is said to be odd or even as the number of arcs in its boundary is odd or even. Prove that the total number of *odd* countries which are colored either one or the other of two specified colors, say A and B, is always an even number.

Solution:

The question appears to be very difficult because, even if we know that a country is colored A, we don't know if it should be included in our count until we find out whether it is odd or even. The following brilliant notion comes to our rescue. Each vertex V is enclosed in a little triangle as shown (FIG. 32) (the vertex V and a short piece of each arc at V are rubbed out and the loose ends are joined to form a triangle). Of course, this adds an extra triangular country at each vertex to give us a new map M'. Although three of the four colors occur at V, there is still the fourth color left for the new triangle. Thus these triangles may be

*AMM, 1966, p. 204, Problem E1756, proposed by J. P. Ballantine, University of Washington, solved by Agnis Kaugars, Kalamazoo College.

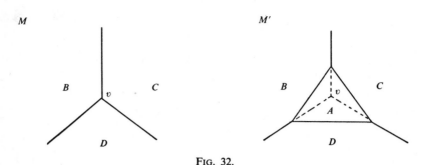

FIG. 32.

colored in keeping with the rule of having adjacent countries different colors. Observe also that there are still exactly three countries at each vertex in M'.

The effect of constructing M' is two-fold. An edge of a new triangle is also a new edge of an existing country, thereby changing it from an odd country to an even one, or vice-versa. Thus, when the new triangle is drawn at a vertex V, the number of odd regions of each of the three colors occurring at V changes by one (either up or down). But since the new triangle, itself, becomes an odd country of the fourth color, we see that the number of odd countries changes by one for each of the four colors. For any pair of colors, then, the total change in the number of odd countries having those colors is either 2, 0, or -2, depending upon whether the individual changes are increases or decreases. In all cases, the total change for any two colors is an even number. Therefore the new and old numbers of odd countries colored A or B have the same *parity*. This persists through the changes at every vertex. Thus, after constructing M', the total number of odd regions colored A or B is odd or even as it was in M to begin with. Hence we can obtain the desired information by considering the map M'.

Secondly, in making M', the number of edges around each country of M is doubled. Thus, in M', all the former countries are even. Only the new triangles are odd. We wish to show, then, that the number of new triangles in M' which are colored A or B is even.

Now, we really don't need to draw the little triangles. Since they are colored with the color that does not occur at V, we obtain an equivalent situation simply by labeling each *vertex* of M with the color not occurring there. Let this be done instead. We will show that the total number of vertices which are thus labeled A or B is even.

To this end, consider an arc which separates countries that are colored C and D. The vertices at the ends of such a "divider" cannot be labeled C or D since these colors occur at each of them (FIG. 33). Therefore the ends of a divider must be labeled A or B, accounting for two such vertices. On the other hand, if a vertex is labeled A or B, then C and D must be included among the other three colors which are used on the countries at the vertex. Consequently, one of the edges at such a vertex must be a divider. We have, then, that both ends of a divider are either A or B, and that the dividers account for all the vertices labeled A or B.

Our argument concludes with the proof that no two dividers share a common endpoint. But this is immediate. Suppose two dividers meet at a vertex labeled A (say). The three countries at this vertex must be colored B, C, and D. But the edge between the countries colored B and C is not a divider; neither is the one between the countries colored B and D. Thus not more than one of the edges at a vertex labeled A can be a divider.

This means that the dividers are separate. The vertices labeled A or B, then, go together in *pairs*, implying that their total number is even.

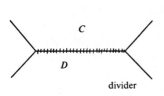

divider

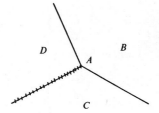

FIG. 33.

CONVEX REGIONS IN THE PLANE*

Prove that every closed convex region in the plane of area π (or more) has two points which are two units apart (FIG. 34).

Solution:

Let polar coordinates be assigned so that the region is tangent to the polar axis at the origin. Suppose the equation of the boundary of the region is $r = f(\theta)$. Then, by the usual formula for area, we have

$$A = \int_0^\pi \frac{1}{2} r^2 d\theta = \frac{1}{2} \int_0^\pi f^2(\theta) d\theta$$

$$= \frac{1}{2} \int_0^{\pi/2} f^2(\theta) d\theta + \frac{1}{2} \int_{\pi/2}^\pi f^2(\theta) d\theta.$$

Changing from θ to $\theta + \pi/2$ in the second integral, we see that it is equal to

$$\frac{1}{2} \int_0^{\pi/2} f^2\left(\theta + \frac{\pi}{2}\right) d\left(\theta + \frac{\pi}{2}\right) = \frac{1}{2} \int_0^{\pi/2} f^2\left(\theta + \frac{\pi}{2}\right) d\theta.$$

Hence the area is

$$A = \frac{1}{2} \int_0^{\pi/2} \left[f^2(\theta) + f^2\left(\theta + \frac{\pi}{2}\right) \right] d\theta.$$

The integrand here is the square of a chord of the region in

*Pi Mu Epsilon, 1956, p. 185, Problem 74, proposed and solved by H. Helfenstein, University of Alberta.

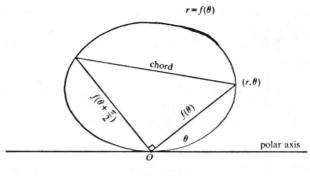

FIG. 34.

question. If no two of its points are as far apart as two units, then all of its chords will be less than two units, with squares less than 4, and we have

$$A < \frac{1}{2}\int_0^{\pi/2} 4\,d\theta = 2\int_0^{\pi/2} d\theta = \pi,$$

a contradiction. The conclusion follows.

We observe that, if no chord exceeds 2, then the area cannot exceed π. Therefore, if the area is known to exceed π, we conclude that some chord must exceed 2.

This work leads to some nice results in the modern subject of combinatorial geometry. In 1911, a posthumous paper of Hermann Minkowski touched off a series of striking results concerning closed convex regions in the plane with the now famous theorem:

If a centrally-symmetric closed convex region of area >4 has its center on a lattice point, then it covers at least 2 additional lattice points.

(A centrally-symmetric figure is one which contains a point O about which a half-turn carries the figure into itself.)

Having discussed this theorem in Mathematical Gems, Vol. 1, p.

47, I will not repeat its proof here. However, a theorem which we are in a position to handle easily is the following, due to Joseph Hammer, University of Sydney (AMM, 1968, p. 157, Mathematical Notes):

If a centrally-symmetric closed convex region R of area $> \pi$ has its center O at a lattice point, then it can be **rotated** *about its center to a position where it covers at least 2 additional lattice points.*

Proof. Since O is the center of R, it bisects every chord of R that goes through it. Therefore, if there is a chord AB through O of length ≥ 2, the chord will extend at least 1 unit on either side of O and, upon being rotated about O, will be carried into a position to cover the two adjacent lattice points (C and D, or E and F) on a lattice line through O (FIG. 35).

Since R has area $> \pi$, we know that some chord AB of R must exceed 2. Thus, if AB goes through O, the conclusion follows. If AB does not go through O, then O is even farther than $AB/2$ from one end or the other, say A. Because R is centrally symmetric, in this case the chord AOA' is even longer than AB, and the proof is complete.

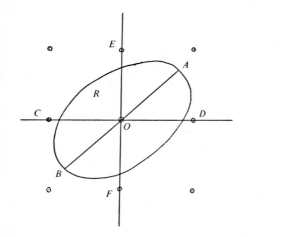
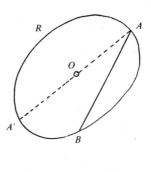

FIG. 35.

It is precisely the same idea which establishes a second theorem in Hammer's article:

If a closed convex region R of area $> \pi/2$ is rotated about an **arbitrary** *point P of the region, then in some position it covers at least one lattice point.*

Proof. If R is subjected to the geometric dilatation $P(\sqrt{2})$ (that is, center P, magnification factor $\sqrt{2}$), its linear dimensions increase by a factor of $\sqrt{2}$ and its area doubles. Since the new area would therefore exceed π, we see that the new region would contain a chord >2. This means that the corresponding chord AB in R has length $>2/\sqrt{2} = \sqrt{2}$. Now, as before, either the point P is the midpoint of this chord AB or it is even farther than $AB/2$ from one of A or B. In any case, one of PA or PB exceeds $\sqrt{2}/2$, say PA. But every point of the plane is at most $\sqrt{2}/2$ from a lattice point (the center C of a lattice square is exactly $\sqrt{2}/2$ from each of the four lattice-point vertices) (Fig. 36). Thus, rotating about P carries the spoke PA to cover the lattice point which is nearest P.

The regions involved in Minkowski's theorem and Hammer's first theorem are required to be centrally symmetric. For a centr-

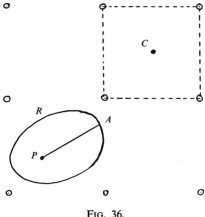

Fig. 36.

ally-symmetric figure, the center of symmetry and the center of gravity coincide. Thus these theorems could be stated equivalently in terms of regions having their centers of gravity placed on a lattice point. If now the requirement of central symmetry be dropped, these theorems make sense but they are untrue. However, to our surprise, they can be salvaged by increasing the area requirement by a factor of $1/8$:

A closed convex region of area $>4\frac{1}{2}$, *with its center of gravity at a lattice point, covers at least* 2 *additional lattice points* (due to E. Ehrhart, [1]);

A closed convex region of area $>9\pi/8$, *with its center of gravity at a lattice point, can be rotated about its center of gravity to a position where it covers at least* 2 *additional lattice points* (due to Hammer, in the afore-mentioned article in the AMM, 1968).

The reader might enjoy the very brief article [2].

References

1. E. Ehrhart, Une généralisation du théorème de Minkowski, Comptes Rendus, 240 (1955), 483–485.

2. P. R. Scott, Lattice Points in Convex Sets, Math. Mag., May 1976, 145–146.

SIMULTANEOUS DIOPHANTINE EQUATIONS*

Solve the following system of equations in natural numbers:

$$a^3 - b^3 - c^3 = 3abc,$$
$$a^2 = 2(b + c).$$

Solution:

Since $3abc$ is positive, a^3 must be greater than either b^3 or c^3, giving

$$b < a \quad \text{and} \quad c < a.$$

Adding gives $b + c < 2a$, and therefore $2(b + c) < 4a$. From the second equation, then, we have

$$a^2 < 4a, \quad \text{and} \quad a < 4.$$

But the second equation also shows that a is an even number. Thus a must be 2, and the lesser b and c must each be 1.

*AMM, 1958, Problem E1266, proposed by D. C. B. Marsh, Colorado School of Mines, solved by Raymond Huck, Marietta College.

A REFLECTED TANGENT*

A and *B* are two circles on the same side of a line *m*. Construct a tangent to *A* which, upon reflection by *m*, is also a tangent to *B* (FIG. 37.)

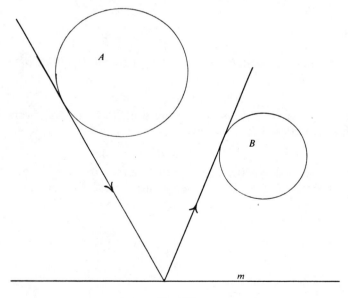

FIG. 37.

*AMM, 1901, p. 144, Problem 152, proposed by Elmer Schuyler, Reading, Pennsylvania, solved by Marcus Baker, Washington D.C.

Solution:

Reflect *B* in the line *m* to get *B'*. Then the common tangents of *A* and *B'* yield the four possible solutions. (Tangents to *B'* reflect in *m* into tangents to *B*.) (FIG. 38)

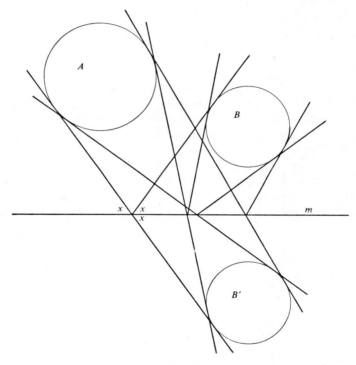

FIG. 38.

THE ELEGANTLY DESTROYED
CHECKERBOARD*

If all the black squares are removed from an ordinary 8×8 checkerboard, the board loses the ability to hold even a single 2×1 domino. However, no matter which black square is returned to the board, it regains this ability. Such a reduced checkerboard is said to be "elegantly destroyed."

Now it is not necessary to take away all the black squares (or all the white ones) in order to destroy the board. Various mixtures of black and white squares suffice. Thirty-two dominos cover a checkerboard in an obvious way (FIG. 39) and, unless each of

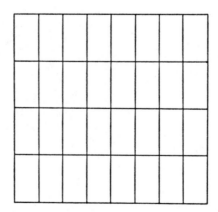

FIG. 39.

*This problem and solution are due to Sidney Penner, Bronx Community College, City University of New York.

these 32 pairs of adjacent squares is ruptured, the board will retain its ability to hold a domino. Therefore, in order to destroy a board, a minimum of 32 squares must be deleted.

On the other hand, it turns out to be possible to remove more than 32 squares in making an elegantly destroyed board. Of course, many wholesale reductions will destroy a board. The trick is to arrange things so that the return of *any* deleted square again renders the board capable of accommodating a domino. In view of this rather stringent constraint, the maximum number of squares that can be deleted in making such a board is surprisingly large. Determine this maximum.

Solution:

The board consisting of the shaded squares in Fig. 40 demonstrates that it is possible for an elegantly destroyed board to retain only 16 squares. We shall show that fewer than 16 squares is impossible by proving that, in every reduction of more than 48 squares, there is at least one square, which we denote X, whose return to the board fails to restore its capability of holding a domino.

If as many as 4 squares were to be left in each quarter of the board, a total of at least 16 squares would remain. In a board with

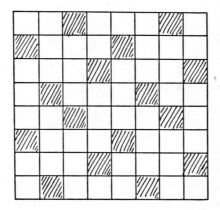

Fig. 40.

fewer than 16 squares, then, there must be some quarter, say the upper left one, which retains no more than 3 of its squares. Consequently, some quarter of this quarter of the board must have been completely demolished. We consider, in turn, the four possible positions A, B, C, D for this vacant quarter of the quarterboard (FIG. 41).

(a) If the vacant quarter is A, then the return of the very corner square X still does not permit the board to accept a domino (FIG. 42).

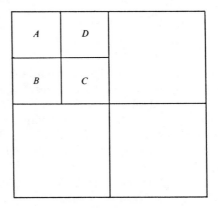

FIG. 41.

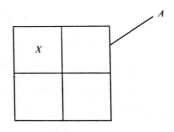

FIG. 42.

(b) Suppose the four squares E, F, G, H of section B are deleted, and suppose the squares around B are labeled as shown in

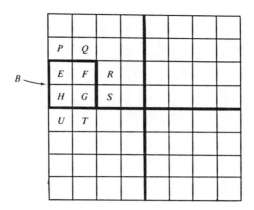

Fig. 43.

FIG. 43. Now, unless the square P is retained, the return of E will not restore the board's capability of accepting a domino, and our search for an impotent square X will be at an end. Suppose, therefore, that P is retained. In this case, the adjacent square Q must have been deleted in order to "destroy" the board.

Around F, then, we have empty squares Q, E, and G, implying that either R is retained or $X \equiv F$ ends our search. Suppose that R is kept. Then the adjacent S must have been removed. Considering G, we achieve a desired impotent square $X \equiv G$ unless T is retained, which necessitates U being deleted. In all this, then, we have either found a suitable X or deduced that the square U is deleted, implying that all three squares which are adjacent to H are missing, giving $X \equiv H$. Therefore an X can be found in any case.

The equivalent quarter D is handled identically.

(c) It remains only to consider the vacant quarter in position C. Let the labeling of FIG. 43 be extended as shown in FIG. 44. The section C consists of the squares R, S, W, V.

Consider first the square R. Since V and S are deleted, either F or Y must be retained or we have $X \equiv R$. Since F and Y are similarly positioned relative to the quarter C, the cases are equivalent. Suppose, for definiteness, that Y is retained. In this case, we

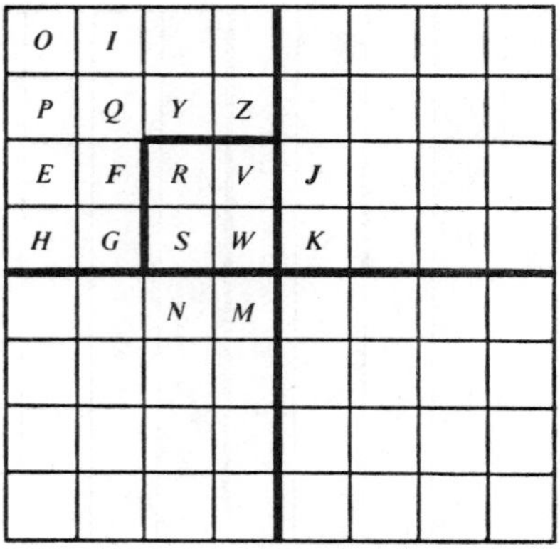

Fig. 44.

can work our way around C, as we did above around B, to discover an impotent square X (V, W, or S) or to find that Y, J, M, and G must be retained and Z, K, N, and F must be deleted. Thus either we obtain X or the two squares Y and G must be retained in the upper left-hand quarter of the board. But recall that, in this quarter-board, at most 3 squares are retained altogether.

If two of these are Y and G, there can be at most one additional square which is retained in this entire quarter of the board.

If the outer corner O is also kept, then P, E, H, and F must be removed, yielding $X \equiv E$. And if O is not retained, we easily complete the argument as follows. If P is kept, then $X \equiv I$. Suppose, then, that P is deleted. Because Y is kept, Q must be deleted. In the event that E is also deleted, we have $X \equiv P$ (since O is gone). If E is retained, making the three retained squares Y, G, and E, then $X \equiv I$.

THE SNOWBALLS*

A boy makes two snowballs, one having twice the diameter of the other. He brings them into a warm room and lets them melt. Since only the surface of a snowball is exposed to the warm air, assume that the rate of melting is proportional to the surface area. When half the volume of the larger one has melted away, how much is left of the smaller one?

Solution:

We shall prove that the assumption that a snowball melts at a rate which is proportional to its surface area leads to the striking result that the radius decreases at a constant rate regardless of its size. As a consequence, both radii will shrink by the same amount.

The volume and surface area, respectively, are $V = 4\pi r^3/3$ and $S = 4\pi r^2$. Letting t denote time, the rate of melting is

$$\frac{dV}{dt} = \frac{4}{3}\pi \cdot 3r^2 \cdot \frac{dr}{dt} = 4\pi r^2 \cdot \frac{dr}{dt}.$$

Since this is proportional to $S = 4\pi r^2$, we have

$$4\pi r^2 \cdot \frac{dr}{dt} = k(4\pi r^2),$$

where k is a constant, giving

$$\frac{dr}{dt} = k,$$

as claimed.

*MM, 1944-45, p. 96, Problem 557, proposed by E. P. Starke, Rutgers University, solved by Frank Hawthorne, New Rochelle, New York.

　　Suppose the radii at the beginning are $2r$ and r. The larger ball, then, has volume

$$V = \frac{4}{3}\pi(2r)^3 = \frac{32}{3}\pi r^3.$$

When half melted, it has volume

$$V = \frac{16}{3}\pi r^3 = \frac{4}{3}\pi(\sqrt[3]{4}\cdot r)^3,$$

implying a radius of $\sqrt[3]{4}\cdot r$. Accordingly, each radius has shrunk by an amount $(2 - \sqrt[3]{4})r$. The smaller, then, still has radius

$$r - (2 - \sqrt[3]{4})r = r(\sqrt[3]{4} - 1).$$

The amount left in the smaller ball is therefore

$$V = \tfrac{4}{3}\pi r^3(\sqrt[3]{4} - 1)^3,$$

which is very nearly 1/5 of its original volume $((\sqrt[3]{4} - 1)^3 = .2027$ approximately).

WRITING THE NUMBERS FROM ONE TO A BILLION*

What is the sum of all the digits used in writing down the numbers from one to a billion?

Solution:

Throwing in zero, we can form one-half a billion pairs of numbers

$$(0; 999,999,999), (1; 999,999,998), (2; 999,999,997), \ldots,$$

$$(499,999,998; 500,000,001), (499,999,999; 500,000,000),$$

the sum of the digits in each pair being $9 \cdot 9 = 81$. Including 1 for the sum of the digits in the neglected 1,000,000,000, the required sum is simply

$$500,000,000(81) + 1 = 40,500,000,001.$$

*Scripta Mathematica, 1950, p. 126, Curiosa 224, by Leo Moser.

ABUTTING, NONOVERLAPPING UNIT SQUARES*

Let a unit square S in the plane be fixed in position. What is the maximum number of nonoverlapping unit squares which can be placed up against S so as to touch S but not overlap it? (FIG. 45.)

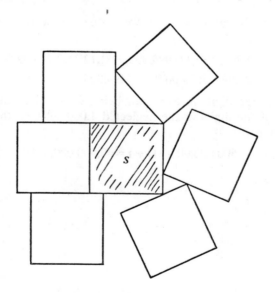

FIG. 45.

*AMM, 1939, p. 20, article "A Lemma on Squares" by J. W. T. Youngs, Ohio State University.

Solution:

A 3×3 checkerboard pattern provides 8 abutting squares and certainly seems to be an arrangement in which the squares are packed as tightly as possible (FIG. 46). Having felt this to be the case for a long time, but utterly unable to prove it, I was surprised and delighted one day to run across J. W. T. Youngs' marvellous proof while browsing through the 1939 volume of the American Mathematical Monthly.

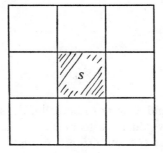

FIG. 46.

From the diagrams (FIG. 47), it is clear that
(i) the distance between the centers of two nonoverlapping unit squares is $\geqslant 1$, and
(ii) the distance between the centers of two abutting unit squares cannot exceed $\sqrt{2}$.
Let A and B denote the centers of two of the squares which abut the fixed square S, with center O (FIG. 48). Let $OA = x$, $OB = y$, and $AB = t$. Then by (i) and (ii) above, we have

$$x, y, t \geqslant 1 \quad \text{and} \quad x, y \leqslant \sqrt{2}.$$

There could possibly be a gap separating the squares with centers A and B, thus allowing t to range $> \sqrt{2}$. Since none of the squares overlap, however, we do have $t \geqslant 1$.

(*i*)   (*ii*)

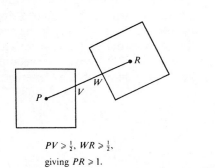

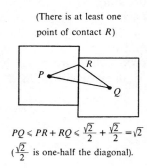

(There is at least one point of contact R)

$PV \geqslant \frac{1}{2}$, $WR \geqslant \frac{1}{2}$, giving $PR \geqslant 1$.

$PQ \leqslant PR + RQ \leqslant \frac{\sqrt{2}}{2} + \frac{\sqrt{2}}{2} = \sqrt{2}$
($\frac{\sqrt{2}}{2}$ is one-half the diagonal).

FIG. 47.

Now, if O, A, and B were to be collinear, say in that order, the distance $OB = y$ would have to be at least 2 units, exceeding its limit of $\sqrt{2}$. Hence OA and OB are *distinct* lines radiating from O (FIG. 48). Consequently, joining O to all the centers of the squares which abut S provides a fan of distinct lines through O, containing a line for each square.

Suppose, now, that OA and OB are *consecutive* lines in the fan and that the angle they make at O is denoted θ, as shown. The law of cosines applied to $\triangle OAB$ yields

$$t^2 = x^2 + y^2 - 2xy \cos\theta,$$

from which

$$\cos\theta = \frac{x^2 + y^2 - t^2}{2xy}.$$

Since $t \geqslant 1$, we have

$$\cos\theta \leqslant \frac{x^2 + y^2 - 1}{2xy},$$

which we denote $f(x,y)$. Recalling that $1 \leqslant x,y \leqslant \sqrt{2}$, we see that

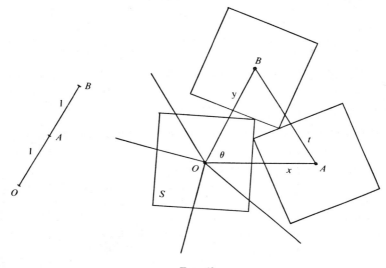

Fig. 48.

$f(x,y)$ is positive. We now proceed to show that, for the x and y under consideration, $f(x,y)$ never exceeds 3/4.

For the reader who is not familiar with partial differentiation, a detailed elementary derivation is given in the Appendix. We have

$$f_x = \frac{2xy(2x) - (x^2+y^2-1)(2y)}{(2xy)^2}$$

$$= \frac{x(2x) - (x^2+y^2-1)}{2x^2y} = \frac{x^2-y^2+1}{2x^2y}.$$

Similarly,

$$f_y = \frac{-x^2+y^2+1}{2xy^2}.$$

For $1 \leqslant x,y \leqslant \sqrt{2}$, it is clear that both f_x and f_y are nonnegative. Therefore the value of f is nondecreasing as x increases and as y increases, implying that $f(\sqrt{2}, \sqrt{2})$ is as great as any $f(x,y)$

under consideration. That is to say

$$f(x,y) \leqslant f(\sqrt{2}, \sqrt{2}) = \frac{2+2-1}{2 \cdot \sqrt{2} \cdot \sqrt{2}} = \frac{3}{4}.$$

We see, therefore, that the angle θ between OA and OB is such that

$$\cos\theta \leqslant \tfrac{3}{4}.$$

However, from a set of tables we learn that $\cos 40° = .76604\ldots$. Thus

$$\cos\theta < \cos 40°.$$

This means that $\theta > 40°$, implying that, in the 360°-sweep of the fan of lines radiating to the centers of the abutting squares from O, there is not enough room for as many as 9 such angles θ, each $> 40°$, giving the maximum number of abutting squares to be 8.

A similar, much easier problem was asked in the American Mathematical Monthly (1962, p. 808) by D. J. Newman and W. E. Weissblum, Yeshiva University:

Six closed circles in the plane are such that none contains the center of another. Prove that they cannot have a point in common.

Solution:

Suppose that some point O is common to all the circles. Let O be joined to the 6 centers to form a fan of lines at O. If two centers A and B were to lie on the same ray through O, then, because each circle contains O, it is evident that one circle must contain the center of another (a contradiction). Thus 6 different lines occur in the fan (FIG. 49).

Let A and B denote two centers such that OA and OB are consecutive lines in the fan. Let r denote the greater of the radii of the circles with centers A and B (their common value if they are equal). Then, because no circle contains the center of another, AB must exceed r. But, since both the circles, centers A and B, contain O, neither of OA and OB can exceed r. Thus, in $\triangle OAB$, AB is definitely the longest side, making the angle AOB greater then

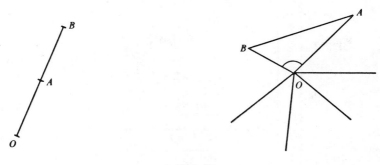

FIG. 49.

either of the other angles. Consequently,

$$\text{angle } AOB > 60°,$$

which means that the fan of 6 lines contains a total of more than $6(60) = 360$ degrees, which is impossible.

Appendix:

The graph of the function $z = f(x,y) = (x^2 + y^2 - 1)/(2xy)$ is a 3-dimensional surface (FIG. 50). We are interested only in the part G which lies directly above the square $1 \leqslant x, \, y \leqslant \sqrt{2}$ in the xy-plane. Let $K(a,b)$ denote any point in this square and let $P(a,b,z)$ denote the point on G directly above K. Consider the line L of points in the square which have the same y-coordinate as K. The plane π through L which is perpendicular to the xy-plane intersects G in some curve C. Being on G, all the points of C have coordinates which are related by

$$z = \frac{x^2 + y^2 - 1}{2xy}.$$

However, lying in the plane π, they all have $y = b$. Hence, for the curve C, we have

$$z = \frac{x^2 + b^2 - 1}{2bx}.$$

This is a function of the single variable x and the slope of the

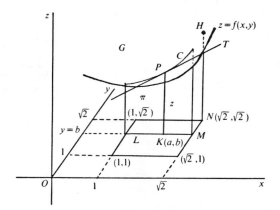

FIG. 50.

tangent T to the curve C at the point P on it is simply the value of dz/dx at the point P. We have, after simplification,

$$\frac{dz}{dx} = \frac{x^2 - b^2 + 1}{2bx^2}.$$

Since $x \geqslant 1$, then $x^2 + 1 \geqslant 2$, and since $b \leqslant \sqrt{2}$, the numerator $x^2 - b^2 + 1 \geqslant 0$. Because the denominator is positive, we conclude that the slope of the tangent at *each* point of C is nonnegative. That is to say, as one moves along C in the direction of increasing x, the value of z does not decrease. Hence the endpoint M of L provides a point on G which is at least as far above the xy-plane as any point on G which is directly above L.

By running the line L and the plane π parallel to the y-axis, we similarly obtain a curve C', given by

$$z = \frac{a^2 + y^2 - 1}{2ay}.$$

For points on C' we have

$$\frac{dz}{dy} = \frac{-a^2 + y^2 + 1}{2ay^2},$$

which is also seen to be nonnegative for the y and a in question. Thus, one remains at least as far above the xy-plane by moving on G so as to increase y while keeping x constant. We conclude that, beginning at any point $K(a,b)$ in the square, one at no time lies under a part of G which is closer to the xy-plane if one moves along L to M and then along the edge of the square to the corner N. Consequently, no point of G is farther from the xy-plane than the point H which is directly above N. Thus the point $N(x,y)$, which is $N(\sqrt{2}, \sqrt{2})$, provides the maximum value of $f(x,y)$ for (x,y) in the square in question. And this maximum value is

$$\frac{2+2-1}{2\cdot\sqrt{2}\cdot\sqrt{2}} = \frac{3}{4}.$$

A DIOPHANTINE EQUATION*

Let a, b, c, d be integers, with $a \neq 0$. Prove that, so long as $bc - ad \neq 0$, the equation

$$axy + bx + cy + d = 0$$

has only a finite number of solutions (x, y) in integers.

Solution:

Multiplying by a, we get

$$a^2 xy + abx + acy + ad = 0,$$

and

$$(ax + c)(ay + b) = bc - ad.$$

This means that each of $ax + c$ and $ay + b$ is a divisor of $bc - ad$. Since $bc - ad \neq 0$, it has only a finite number of divisors, implying a finite number of possible values for $ax + c$ and $ay + b$, and a like number of x and y.

We observe that the graph of $axy + bx + cy + d = 0$ is a hyperbola. If $bc - ad = 0$, the equation reduces to

$$(ax + c)(ay + b) = 0,$$

and the hyperbola degenerates into a pair of straight lines parallel to the axes. Clearly the integral solutions (x, y) of the equation correspond to the lattice points on the graph. Thus, even for $bc - ad = 0$, there is only a finite number of integral solutions

*AMM, 1964, p. 794, Problem E1631, proposed by Roy Feinman, Rutgers University, solved by Richmond G. Albert, West Newton, Massachusetts.

unless, in addition, a divides either c or b, which places a line of the degenerate curve in coincidence with a lattice line (parallel to a coordinate axis).

The following neat solution to this problem was pointed out by my colleague Leroy Dickey. The hyperbola given by $axy + bx + cy + d = 0$ happens to have asymptotes which are parallel to the axes of coordinates. This may be seen to be the case by observing that the *translation*

$$x = X - \frac{c}{a}, \qquad y = Y - \frac{b}{a}$$

transforms the equation into

$$a\left(X - \frac{c}{a}\right)\left(Y - \frac{b}{a}\right) + b\left(X - \frac{c}{a}\right) + c\left(Y - \frac{b}{a}\right) + d = 0,$$

$$aXY + k = 0, \qquad \text{where } k \text{ is a constant,}$$

yielding $XY = $ (a constant), the standard equation of a hyperbola referred to its asymptotes as axes. Because of this, as the curve approaches its asymptote, there comes a time when it crosses a lattice line for the last time, thereafter to remain in a channel between a pair of adjacent lattice lines. This is true even when an asymptote coincides with a lattice line, for a hyperbola does not cross its own asymptote. Once in such a channel, the infinite tail of the curve encounters no more lattice points (FIG. 51). Consequently, the lattice points which the hyperbola does encounter all lie in a finite part of the plane and therefore must be finite in number.

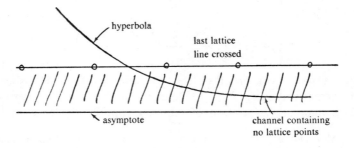

FIG. 51.

THE FIBONACCI SEQUENCE*

The sequence of natural numbers

$$\{f_n\}:\ 1,1,2,3,5,8,13,21,34,55,\dots,$$

where

$$f_1=f_2=1 \text{ and } f_n=f_{n-1}+f_{n-2} \quad \text{for} \quad n>2,$$

is called the Fibonacci sequence and is one of the most popular sequences in mathematics. In fact, it is so rich in properties and generalizations that an entire journal, The Fibonacci Quarterly, is dedicated to the investigation of this sequence and topics related to it.

Our problem is simply how many terms of the Fibonacci sequence are less than or equal to a given natural number N?

Solution:

It has been known for well over a century that the nth term of the Fibonacci sequence is

$$f_n = \frac{1}{\sqrt{5}}\left[\left(\frac{1+\sqrt{5}}{2}\right)^n - \left(\frac{1-\sqrt{5}}{2}\right)^n\right].$$

(For a very neat proof of this formula see Ross Honsberger: Mathematical Gems, Vol. 1, Dolciani Mathematical Expositions, Mathematical Association of America, pp. 171-2.) Now, $\sqrt{5}$ is

*AMM, 1964, p. 798, Problem E1636, proposed by J. D. Cloud, North American Aviation, Inc., solved by William D. Jackson, State University College, Oswego, New York.

approximately 2.2, making

$$\frac{1-\sqrt{5}}{2} = -.6 \text{ approximately.}$$

Thus $((1 - \sqrt{5})/2)^n$ is positive or negative as n is even or odd and, for all $n = 1, 2, \ldots,$ the quantity

$$\frac{1}{\sqrt{5}} \left(\frac{1-\sqrt{5}}{2} \right)^n$$

has magnitude $< 1/2$. But f_n is a whole number. Since

$$\left| -\frac{1}{\sqrt{5}} \left(\frac{1-\sqrt{5}}{2} \right)^n \right| < \frac{1}{2},$$

the formula shows that f_n must be the natural number which is nearest to

$$\frac{1}{\sqrt{5}} \left(\frac{1+\sqrt{5}}{2} \right)^n.$$

It is also clear that

$$\frac{1}{\sqrt{5}} \left(\frac{1+\sqrt{5}}{2} \right)^n$$

is closer than $1/2$ to the integer which is nearest to it. Therefore it can never be true that

$$\frac{1}{\sqrt{5}} \left(\frac{1+\sqrt{5}}{2} \right)^n = N + \tfrac{1}{2}.$$

For a Fibonacci number $f_n \leq N$, then, we must have

$$\frac{1}{\sqrt{5}} \left(\frac{1+\sqrt{5}}{2} \right)^n < N + \tfrac{1}{2}$$

(lest the nearest integer, f_n, exceed N). Conversely, it is clear that, if

$$\frac{1}{\sqrt{5}} \left(\frac{1+\sqrt{5}}{2} \right)^n < N + \tfrac{1}{2},$$

the nearest integer must be $\leq N$. Therefore we have $f_n \leq N$ if and only if

$$\frac{1}{\sqrt{5}}\left(\frac{1+\sqrt{5}}{2}\right)^{n} < N + \frac{1}{2},$$

$$\left(\frac{1+\sqrt{5}}{2}\right)^{n} < \left(N + \frac{1}{2}\right)\sqrt{5},$$

$$n \cdot \log\left(\frac{1+\sqrt{5}}{2}\right) < \log\left\{\left(N + \frac{1}{2}\right)\sqrt{5}\right\},$$

$$n < \frac{\log\left\{\left(N + \frac{1}{2}\right)\sqrt{5}\right\}}{\log\left(\frac{1+\sqrt{5}}{2}\right)}.$$

Now, it is easy to see that this ratio of logarithms is never an integer. Suppose, to the contrary, that

$$\frac{\log\left\{\left(N + \frac{1}{2}\right)\sqrt{5}\right\}}{\log\left(\frac{1+\sqrt{5}}{2}\right)} = k, \text{ an integer.}$$

Then

$$\log\left\{\left(N + \frac{1}{2}\right)\sqrt{5}\right\} = k \cdot \log\left(\frac{1+\sqrt{5}}{2}\right) = \log\left(\frac{1+\sqrt{5}}{2}\right)^{k},$$

$$\left(N + \frac{1}{2}\right)\sqrt{5} = \left(\frac{1+\sqrt{5}}{2}\right)^{k}, \text{ and } \frac{1}{\sqrt{5}}\left(\frac{1+\sqrt{5}}{2}\right)^{k} = N + \frac{1}{2},$$

contradicting the result above.

Since n is an integer, we see that the greatest Fibonacci number $\leqslant N$, corresponding to the greatest permissible value of n, is the term for which

$$n = \left[\frac{\log\left\{ \left(N + \frac{1}{2}\right)\sqrt{5} \right\}}{\log\left(\frac{1 + \sqrt{5}}{2}\right)} \right],$$

where $[x]$ denotes the greatest integer $\leqslant x$. Since the first two terms are the same, we see that the number of different values among these Fibonacci numbers is 1 less than this number n.

Consider now the problem of determining the number of ways of making a selection from the numbers $1, 2, 3, \ldots, n$ without taking a pair of consecutive numbers (counting the empty set as a selection).

Solution:

Suppose we let a_{n-1} denote the number of selections for the numbers $1, 2, 3, \ldots, n-1$. Consider, then, the a_n selections from $1, 2, 3, \ldots, n$. Such a selection either contains the number n or it doesn't. If it doesn't, then it is one of the a_{n-1} selections from $1, 2, 3, \ldots, n-1$. If it does, then the number $n-1$ must be avoided (since $n-1$ and n are consecutive), implying the remaining part of the selection is one of the a_{n-2} from $1, 2, 3, \ldots, n-2$. Of course, it is not necessary to select more than just the number n, itself. This corresponds to the selection of the empty set from $1, 2, 3, \ldots, n-2$ (this is why the empty set is included in the count). Thus, altogether, we have

$$a_n = a_{n-1} + a_{n-2}.$$

Since $a_1 = 2$ and $a_2 = 3$ are easily calculated, we see that the sequence proceeds $2, 3, 5, 8, \ldots$, and $a_n = f_{n+2}$, the $(n+2)$th term of the Fibonacci sequence.

AN ERDÖS INEQUALITY*

ON is the radius which is perpendicular to chord AB in a circle, center O, meeting AB at M. P is any point on the major arc AB which is not diametrically opposite N. PM and PN determine, respectively, Q and R on the circle and AB (Fig. 52). Prove that RN is always longer than MQ. (Surprisingly, many people, on a quick guess, choose MQ to be the longer.)

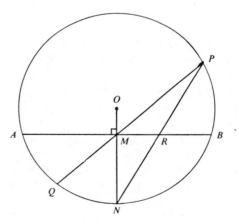

Fig. 52.

Solution I (Paul Erdös):

Let the reflection of PN in the diameter NON' be $P'N$ (Fig. 53). Because AB is perpendicular to ON, this reflection carries R into the intersection R' of $P'N$ and AB. Hence $RN = R'N$.

---

*A geometric inequality of Paul Erdös, received by personal communication, 1975.

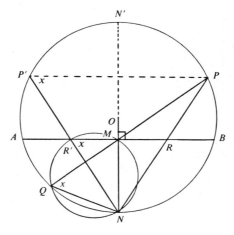

Fig. 53.

Now PP' and AB are both perpendicular to diameter NON'. Hence they are parallel, giving equal corresponding angles $NR'M$ and $NP'P$. But, in the same segment we have $\angle NP'P = \angle NQP$. Thus $\angle NR'M = \angle NQM$, implying $QNMR'$, is cyclic.

Since $R'N$ subtends a right angle at the circumference of this circle (at M), it is a diameter. However, the angle subtended at the circumference by the chord QM, namely $\angle MNQ$, is seen *not* to be a right angle (since NN' is a diameter of the given circle, $\angle NQN'$ is a right angle, implying, in $\triangle QNN'$, that $\angle QNM$ is not). Consequently, the chord QM is less than the diameter $R'N = RN$.

Solution II (Peter Crippen, Scarborough, Ontario):

Let x and y, respectively, denote $\angle P$ and $\angle PMR$ (FIG. 54). Then the angle QON at the center is $2x$, the base angles in isosceles triangle OQN are each $90° - x$, and the other angles are easily calculated to have the sizes shown in the figure. Applying the law of sines to $\triangle MQN$ we have

$$\frac{QM}{\sin(90-x)} = \frac{MN}{\sin(x+y)},$$

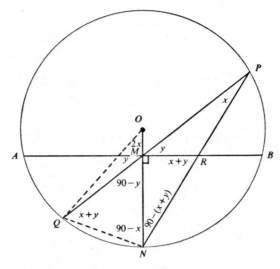

FIG. 54.

and applying it to $\triangle MNR$ we get

$$\frac{MN}{\sin(x+y)} = \frac{RN}{\sin 90}.$$

Thus

$$\frac{QM}{\sin(90-x)} = \frac{RN}{1}.$$

Since $\sin(90-x) < 1$, it follows that $QM < RN$.

A splendid solution was published by the promising high school student Mark Kleiman, New York City, in Math. Mag., 49 (1976), 217–218.

SHARING LATTICE POINTS*

The segment which joins $A(p,0)$ to $B(0,p)$ passes through the $p-1$ lattice points $(1,p-1),(2,p-2),\ldots,(p-1,1)$. The $p-1$ lines from these points to the origin O divide $\triangle OAB$ into p little triangles. Clearly, the two outer little triangles, each having a side along an axis, contain no lattice points of the plane in their interiors. If p is a prime number, it is also true that none of the partitioning segments from the origin contain any lattice points (FIG. 55). Prove that, for p a prime, all the lattice points inside $\triangle OAB$ are in the interiors of the $p-2$ inner little triangles and that they are *shared equally* by them.

Solution:

Let $C(a,b)$ denote a lattice point inside $\triangle OAB$. The slope of the line OC is b/a. If C were to lie on a dividing segment, say the one to $(i,p-i)$ $(1\leqslant i\leqslant p-1)$, the slope of OC would also be $(p-i)/i$. Then $b/a=(p-i)/i$. But, since $i<p$ and p is a prime, i and $p-i$ are relatively prime. However, C is closer to O than $(i,p-i)$. Thus a and b are less, respectively, than i and $p-i$. The equation $b/a=(p-i)/i$, then, indicates that the fraction $(p-i)/i$ is not in its lowest terms, but bears reduction to b/a. This contradicts the fact that i and $p-i$ are relatively prime, and we conclude that all the lattice points inside $\triangle OAB$ occur in the interiors of the $p-2$ inner triangles.

---

*AMM, 1961, p. 806, Problem E1455, proposed by M. T. L. Bizley, London, England, solved by C. M. Superko, Michigan College of Mining and Technology.

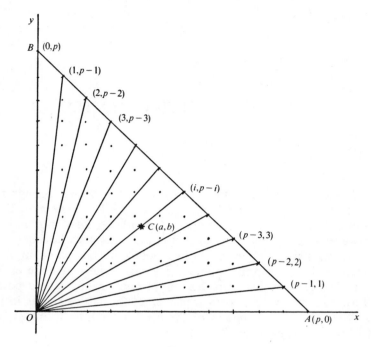

FIG. 55.

Observe that AB is divided into equal parts by the lattice points $(i, p - i)$. This implies that all the little triangles have the same area. Since each vertex is a lattice point, we may use Pick's theorem to determine their area.

PICK'S THEOREM. *The area of a polygon whose vertices are lattice points (and which does not cross itself) is given by*

$$q + \frac{p}{2} - 1,$$

where q denotes the number of lattice points inside the polygon and p denotes the number of lattice points on the boundary (this includes the vertices and any other lattice points on the sides).(See the reference for a beautiful proof of Pick's theorem.)

Because there is no lattice point between $(i, p - i)$ and $(i + 1, p - i - 1)$, we find the area of each inner little triangle to be

$$q + \frac{3}{2} - 1.$$

Since this is the same for all inner triangles, q must be the same for them all, showing that they share the interior lattice points of $\triangle OAB$ equally.

It is easy to compute the value of q. Each little triangle has area $(1/p)(\triangle OAB)$, which is

$$\frac{1}{p}\left(\frac{1}{2} \cdot p \cdot p\right) = \frac{p}{2}.$$

Thus

$$q + \frac{3}{2} - 1 = \frac{p}{2}, \qquad \text{giving } q = \frac{p - 1}{2}.$$

Reference

Ross Honsberger, Ingenuity in Mathematics, vol. 23, New Mathematical Library, Math. Assoc. of America, 27–31.

PERFECT NUMBERS*

The ancient Greeks discovered that some natural numbers n have the remarkable property that the sum of the proper divisors of n is just the number n, itself. For example, $n = 28$ yields

$$1 + 2 + 4 + 7 + 14 = 28.$$

Such numbers they called "perfect." In terms of the arithmetic function $\sigma(n)$, which denotes the sum of all the positive divisors of n (including n, itself), n is perfect if $\sigma(n) = 2n$. Perfect numbers seem to be very rare. The first five are 6, 28, 496, 8128, 33550336, and in 1976 a total of only 24 perfect numbers are known, the greatest being $2^{19936}(2^{19937} - 1)$, containing over 6000 digits.

In the eighteenth century, Euler proved that every *even* perfect number m can be expressed in the form

$$m = 2^{n-1}(2^n - 1), \qquad \text{where } 2^n - 1 \text{ is a prime number.}$$

(For Leonard Eugene Dickson's marvellous proof of this result (1911), see my book *Ingenuity in Mathematics*, vol. 23, in the New Mathematical Library Series, Mathematical Association of America, page 113 ff. The following two easy results (needed below) are also derived there:

  (i) $\sigma(n)$ is multiplicative, i.e., $\sigma(ab) = \sigma(a) \cdot \sigma(b)$, provided a and b are relatively prime.

  (ii) If $2^n - 1$ is a prime number, then so is the exponent n (page 124).)

---

*AMM, 1975, page 1015, Problem E2500, proposed by Richard Herr, University Park, Pa., solved by M. G. Greening, University of New South Wales, Australia.

In the light of this background, the present problem challenges us to find all the perfect numbers n for which $\sigma[\sigma(n)]$ is also a perfect number.

Solution:

First of all, suppose that n is an odd perfect number. Then $\sigma(n) = 2n$, where 2 and n are relatively prime, giving

$$\sigma[\sigma(n)] = \sigma(2n) = \sigma(2) \cdot \sigma(n) = 3 \cdot \sigma(n) = 3(2n) = 6n.$$

Clearly $6n$ is even, and if it is also a perfect number, then, for some prime p, we must have

$$6n = 2^{p-1}(2^p - 1)$$

(p must be a prime because $2^p - 1$ is a prime). However, because n is odd, $6n$ contains merely a single factor 2 (in the 6). This implies that

$$2^{p-1} = 2^1, \text{ making } p = 2.$$

Accordingly,

$$6n = 2(2^2 - 1) = 6, \text{ implying } n = 1.$$

But 1 is not an odd perfect number, and we have a contradiction. Thus there are no odd perfect numbers n such that $\sigma[\sigma(n)]$ is also a perfect number.

Suppose, then, that n is an even perfect number. As such, for some prime p, it can be written

$$n = 2^{p-1}(2^p - 1), \text{ where } 2^p - 1 \text{ is a prime number.}$$

In this case, 2 and $2^p - 1$ are relatively prime, and we have

$$\sigma[\sigma(n)] = \sigma(2n) = \sigma[2^p(2^p - 1)] = \sigma(2^p) \cdot \sigma(2^p - 1)$$
$$= (2^{p+1} - 1)[(2^p - 1) + 1] \text{ (since } 2^p - 1 \text{ is a prime)}$$
$$= 2^p(2^{p+1} - 1).$$

This is clearly even, and if it is also a perfect number, it is already in the Eulerian form for such a number, implying that the exponent $p + 1$ must be a prime number. Thus p and $p + 1$ must both be prime numbers and, since they are consecutive, they can only be 2 and 3. Accordingly,

$$n = 2^{p-1}(2^p - 1) = 2(2^2 - 1) = 6$$

and

$$\sigma[\sigma(n)] = \sigma[\sigma(6)] = \sigma(12) = 28.$$

Therefore $n = 6$ is the only solution.

THE SIDES OF A QUADRILATERAL*

Prove that if the lengths of the sides of a quadrilateral are integers (relative to some unit of length), and if the length of each side divides the sum of the lengths of the other three sides, then some two of the sides are equal.

Solution:

Suppose that no two of the sides are equal. Then let the lengths of the sides be denoted $s_1 > s_2 > s_3 > s_4$ and let p denote the perimeter. We are given that $s_i | p - s_i$ for $i = 1, 2, 3, 4$. In this case, we have also that $s_i | p$. But how often can s_i divide into p?

Consider the case of s_1, the length of the longest side. The sum of any three sides of a quadrilateral always exceeds the fourth side (the three sides constitute the "long way" around the quadrilateral from one end of the fourth side to the other). Thus s_1 cannot be as great as one-half the perimeter: $s_1 < p/2$. However, s_1 must exceed one-quarter of the perimeter since it is the length of the longest side (otherwise all four sides would not add up to p). Hence

$$\frac{p}{4} < s_1 < \frac{p}{2}.$$

This implies that s_1 divides into p more often than twice but not as many as four times. Consequently, it must divide into p three times, and we have $s_1 = p/3$.

*MM, Problem Q315, proposed by D. L. Silverman.

Because $s_2 < s_1$, s_2 must divide into p more often than s_1 does, implying that $p/s_2 \geqslant 4$, and $s_2 \leqslant p/4$. Again, since $s_3 < s_2$, we have $p/s_3 \geqslant 5$, and $s_3 \leqslant p/5$. Similarly, $s_4 \leqslant p/6$. Therefore

$$p = s_1 + s_2 + s_3 + s_4$$

$$\leqslant \frac{p}{3} + \frac{p}{4} + \frac{p}{5} + \frac{p}{6} = \frac{57}{60} p < p,$$

a contradiction. Hence some two sides of the quadrilateral must be equal.

PRIMES IN ARITHMETIC PROGRESSION*

Show that, in any arithmetic progression of natural numbers with common difference less than 2000, it is impossible for as many as 12 consecutive terms all to be prime numbers.

Solution:

Suppose that the n consecutive terms of an arithmetic progression

$$a, a+d, a+2d,\ldots,a+(n-1)d$$

are all prime numbers, $n \geqslant 3$. If $a < n$, one of these terms would be $a + ad = a(1+d)$, which is not a prime since each of a and $1+d$ exceeds 1. Therefore we must have $a \geqslant n \geqslant 3$.

Let p denote a prime number $< n$, and suppose that p does *not* divide d. Consider, then, the first p terms

$$a, a+d, a+2d,\ldots,a+(p-1)d.$$

Let $r_0, r_1,\ldots,r_{p-1}$ denote the p remainders obtained by dividing these terms by p. Since each term is prime, and $p < n \leqslant a$ (the least term), we see that p does not divide any of these terms and that no remainder is 0. Because these p remainders are drawn only from the $p-1$ nonzero remainders $1, 2,\ldots,p-1$, the Pigeon-hole Principle shows that some two of them are equal, say $r_i = r_j$ ($i \neq j$). This implies

$$a + id \equiv a + jd \pmod{p},$$

*AMM, 1934, p. 519, Problem E83, proposed by Morgan Ward, California Institute of Technology, solved by E. P. Starke, Rutgers University.

and

$$(i-j)d \equiv 0 \ (\mathrm{mod}\, p).$$

That is to say, p divides $(i-j)d$. But p is a prime and it does not divide d. Therefore we must have $p|i-j$. However, since i and j are both positive integers less than p, the only way for $p|i-j$ to be possible is for $i-j=0$, leading to the contradiction $i=j$. Thus a prime $p<n$ must divide d.

A string of $n=12$ consecutive prime terms, then, would require d to be divisible by every prime less than 12, namely 2, 3, 5, 7, and 11. Accordingly, d would have to be a multiple of $2\cdot3\cdot5\cdot7\cdot11 = 2310$, which exceeds 2000, and the conclusion follows.

W. Sierpinski's marvellous book *Theory of Numbers* (1964) contains much interesting material on this topic (pp. 121–125). For example, the 10 consecutive terms in the arithmetic progression

$$199, 409, 619, \ldots, 199+9(210)$$

are all prime numbers; also the 13 terms

$$4943 + k(60060), \ k=0,1,\ldots,12 \quad \text{are all primes.}$$

In order to have $n=5$ consecutive terms of an arithmetic progression prime numbers, it is necessary that 2 and 3 divide the common difference d. Therefore d must be a multiple of 6. For $d=6$, we have the example

$$5, \quad 11, \quad 17, \quad 23, \quad 29.$$

However, this is the one and only time that 5 consecutive terms are primes in any arithmetic progression with $d=6$. The reason for this is that any 5 consecutive terms in such a progression

$$a, \quad a+6, \quad a+2\cdot6, \quad a+3\cdot6, \quad a+4\cdot6$$

must contain a multiple of 5. We have $a+i\cdot6 \equiv a+i \ (\mathrm{mod}\,5)$. No matter what a is congruent to modulo 5, one of the values $i=0,1,2,3,4$ makes $a+i \equiv 0 \ (\mathrm{mod}\,5)$. For prime terms, then, one must be 5, itself. In fact, 5 must be the very first term (since $5-6=-1$ could not precede 5) and $5,11,17,23,29$ results.

ON CEVIANS*

Suppose BC is the longest side of $\triangle ABC$. Let a point O be chosen anywhere inside the triangle and let AO, BO, CO cut the opposite sides in A', B', C', respectively. Prove that

$$OA' + OB' + OC' < BC.$$

Solution:

A segment which runs through a triangle from a vertex to the opposite side is called a "cevian". Clearly, a cevian is shorter in length than the longer of the two sides which meet at the vertex from which it is drawn. Accordingly, the longest side of a triangle exceeds all its cevians. Hence BC is longer than each of AA', BB', CC'.

Let OX and OY be parallel, respectively, to AB and AC to give $\triangle OXY$ similar to $\triangle ABC$ (FIG. 56). Since BC is the longest side of $\triangle ABC$, the corresponding side XY is the longest side of $\triangle OXY$. Thus $XY > $ cevian OA'.

Let XS and YT be parallel, respectively, to CC' and BB'. Then $\triangle BXS$ is similar to $\triangle BCC'$. Clearly BC is the longest side of $\triangle BCC'$, and the corresponding BX is the longest side of $\triangle BXS$. Hence

$$BX > SX = OC' \text{ (in parallelogram } C'SXO\text{)}.$$

Similarly,

$$YC > YT = OB'.$$

*AMM, 1937, p. 400, Problem 3746, and 1940, p. 575, Problem 3848, both proposed and solved by Paul Erdös.

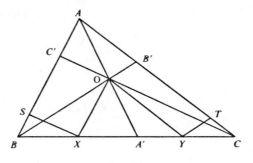

FIG. 56.

Adding, we have

$$OA' + OB' + OC' < XY + YC + BX = BC.$$

It has just been shown that $OA' + OB' + OC' <$ the longest side BC. Suppose that AA' is maximal among the three cevians AA', BB', CC'. Prove, then, the stronger result

$$OA' + OB' + OC' \leqslant AA'.$$

Solution:

Suppose

$$\frac{OA'}{AA'} = x, \ \frac{OB'}{BB'} = y, \ and \ \frac{OC'}{CC'} = z.$$

Let AD and OE be drawn perpendicular to BC (FIG. 57). Then $\triangle ADA'$ and $\triangle OEA'$ are similar, and we have

$$\frac{OE}{AD} = \frac{OA'}{AA'} = x.$$

Now

$$\frac{\triangle OBC}{\triangle ABC} = \frac{\frac{1}{2} \cdot BC \cdot OE}{\frac{1}{2} \cdot BC \cdot AD} = \frac{OE}{AD} = x,$$

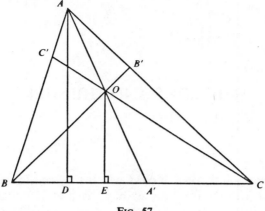

Fig. 57.

giving $\triangle OBC = x(\triangle ABC)$. In a similar way, $\triangle OCA = y(\triangle ABC)$, and $\triangle OAB = z(\triangle ABC)$. Because $\triangle ABC = \triangle OBC + \triangle OCA + \triangle OAB = (x+y+z)(\triangle ABC)$, we obtain $x+y+z=1$.

Then we have

$$OA' + OB' + OC' = x \cdot AA' + y \cdot BB' + z \cdot CC'$$
$$\leqslant x \cdot AA' + y \cdot AA' + z \cdot AA'$$
$$= (x+y+z) \cdot AA'$$
$$= AA'.$$

We observe that the equality holds only if

$$AA' = BB' = CC'.$$

THE COWS AND THE SHEEP*

Two men jointly own x cows which they sell for x dollars per head, and with the returns they buy sheep at 12 dollars per head. As their income from the cows is not divisible by 12, they purchase a lamb with the remainder. Later they divided the flock so that each man had the same number of animals. The man with the lamb was therefore somewhat short-changed. To even things up, the other man gave him his harmonica. How much was the harmonica worth?

Solution:

The income from the sale is x^2 dollars. If x were divisible by 6, then x^2 would be divisible by 36, and hence by 12. Since this is not so, x cannot be a multiple of 6. In this case, $x = 12k + r$, where $|r| = 1, 2, 3, 4$ or 5. Accordingly,

$$\frac{x^2}{12} = \frac{(12k + r)^2}{12} = \frac{144k^2 + 24k + r^2}{12} = 12k^2 + 2kr + \frac{r^2}{12}.$$

Now, since each man has the same number of animals, the total number of animals must be even, implying an odd number of sheep and one lamb. Therefore the quotient provided by $x^2/12$, which is the number of sheep they could buy with x^2 dollars, must be an odd number. But $12k^2 + 2kr$ is an even number. Therefore

---

*AMM, 1930, p. 162, Problem 3379, proposed by J. H. Neelley, and T. L. Smith, Carnegie Institute of Technology, solved by P. S. Ganesa Sastri, Trichinopoly, S. India.

112

$r^2/12$ must provide an *odd* contribution toward this quotient. In order to do so, r^2 must exceed 12, implying $|r|=4$ or 5. For $|r|=5$, we have

$$\frac{r^2}{12} = \frac{25}{12} = 2 + \frac{1}{12},$$

providing an even contribution to the quotient. Hence $|r|$ must be 4, and $r^2 = 16$. Thus

$$\frac{r^2}{12} = \frac{16}{12} = 1 + \frac{4}{12},$$

and the remainder is 4 (not -4). Therefore a lamb cost 4 dollars. Consequently, one man got a 4-dollar lamb and the other a 12-dollar sheep. A 4-dollar harmonica would bring each to the 8-dollar level.

A SEQUENCE OF SQUARES*

Show that each term of the sequence
$$49, 4489, 444889, 44448889, \ldots, \underbrace{44\ldots4}_{n}\underbrace{88\ldots8}_{n}9, \ldots,$$
is a perfect square.

Solution:

The general term is
$$T = \underbrace{44\ldots4}_{n}\underbrace{88\ldots8}_{n}9 = 9 + 8 \cdot 10 + 8 \cdot 10^2 + \cdots + 8 \cdot 10^n$$
$$+ 4 \cdot 10^{n+1} + 4 \cdot 10^{n+2} + \cdots + 4 \cdot 10^{2n+1}.$$

Splitting the 9 into $1 + 4 + 4$ and each 8 into $4 + 4$, we have

$$T = 1 + 4(1 + 10 + 10^2 + \cdots + 10^n) + 4(1 + 10 + \cdots + 10^{2n+1})$$
$$= 1 + 4 \cdot \frac{10^{n+1} - 1}{9} + 4 \cdot \frac{10^{2n+2} - 1}{9}$$
$$= \frac{4 \cdot 10^{2n+2} + 4 \cdot 10^{n+1} + 1}{9} = \left(\frac{2 \cdot 10^{n+1} + 1}{3}\right)^2.$$

This is always the square of an integer because $2 \cdot 10^{n+1} + 1$ can be seen to be divisible by 3 from the fact that the sum of its digits is $2 + 1 = 3$.

*Stanford Competitive Mathematics Examination; AMM, 1973, p. 369-solution not published; solution given here is by Ivan Niven, University of Oregon, and, independently, by Brian Lapcevic, Toronto, Ontario.

THE INSCRIBED DECAGON*

The problem of inscribing regular polygons in a circle has been of keen interest to mathematicians from the time of the Greeks. Euclid wrote on the subject and, in particular, gave a very nice method for inscribing a decagon. We begin by showing that the side x of the regular decagon inscribed in a circle of radius r is

$$x = \frac{r}{2}(\sqrt{5} - 1).$$

Observe that the angle subtended at the center O by the side $AB = x$ is 36° (FIG. 58). The triangle OAB is therefore isosceles with base angles of 72°. Let BC bisect $\angle B$. Then triangles ABC and OBC are both isosceles and $x = AB = BC = OC$. Accordingly, $AC = r - x$.

Consider the circle which circumscribes $\triangle OBC$. Since $\angle ABC$ is the same as that subtended by the chord BC at O on the circumference, AB is tangent to the circle. Consequently, from tangent AB and secant ACO we obtain

$$x^2 = AO \cdot AC = r(r - x) = r^2 - rx,$$
$$x^2 + rx - r^2 = 0,$$
$$x = \frac{-r \pm \sqrt{5r^2}}{2}.$$

Since x is positive, we have

$$x = \frac{r}{2}(\sqrt{5} - 1),$$

as claimed.

*MM, 1953, p. 52, Problem 155, proposed by Leon Bankoff, Los Angeles, California, solved by Daniel Weiner, Wright Junior College, Chicago, Illinois.

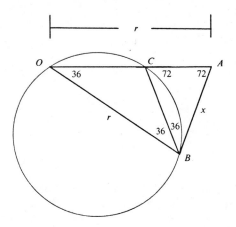

FIG. 58.

Now let us turn to Dr. Bankoff's remarkable problem. We begin with an equilateral triangle *KAB*. Six equal circles are placed around the triangle, three to touch the sides at their midpoints and three to pass through the vertices, the centers lying on extensions of the angle bisectors. The circles are enlarged continuously at the same rate until they grow large enough to touch each other and form a ring of six equal circles of radius r around the fixed triangle (FIG. 59). Then, to our astonishment, the radius x of the circle which is inscribed in the triangle *KAB* is the side of the regular

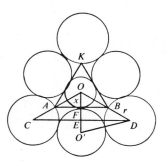

FIG. 59.

decagon in each of the circles in the ring. (Isn't it amazing that anyone would ever make such a discovery!)

Solution:

It is not difficult to see that the inradius x of an equilateral triangle is one-third an altitude (which is also a median and an angle bisector). Referring to the diagram, we have

$$OB = \tfrac{2}{3}(an\ altitude) = 2x.$$

Clearly AB and CD are parallel. Therefore

$$\frac{EF}{OF} = \frac{DB}{BO}, \ \frac{EF}{x} = \frac{r}{2x}, \text{ giving } EF = \frac{r}{2}.$$

Then $O'E = r/2$, the other half of radius $O'F$.
Now, from right-triangle OED, we have

$$ED^2 = OD^2 - OE^2$$
$$= (2x+r)^2 - \left(x + \frac{r}{2}\right)^2$$
$$= 3x^2 + 3rx + \frac{3r^2}{4}.$$

Then right triangle $O'ED$ gives

$$O'D^2 = ED^2 + O'E^2,$$
$$4r^2 = \left(3x^2 + 3rx + \frac{3r^2}{4}\right) + \frac{r^2}{4},$$
$$r^2 = x^2 + rx,$$
$$x^2 + rx - r^2 = 0,$$

yielding, as above,

$$x = \frac{r}{2}(\sqrt{5} - 1).$$

RED AND BLUE DOTS*

Consider a square array of dots, colored red or blue, with 20 rows and 20 columns. Whenever two dots of the same color are adjacent in the same row or column, they are joined by a segment of their common color; adjacent dots of unlike colors are joined by a black segment. There are 219 red dots, 39 of them on the border of the array, none at the corners. There are 237 black segments. How many blue segments are there?

Solution:

There are 19 segments in each of 20 rows, giving $19 \cdot 20 = 380$ horizontal segments. There is the same number of vertical segments, giving a total of 760 altogether. Since 237 are black, the other 523 are either red or blue.

Let r denote the number of red segments and let us count up the number of times a red dot is the endpoint of a segment. Each black segment has one red end, and each red segment has both ends red, giving a total of

$$237 + 2r \text{ red ends.}$$

But, the 39 red dots on the border are each the end of 3 segments, and each of the remaining 180 red dots in the interior of the array is the end of 4 segments. Thus the total number of times a red dot is the end of a segment is

$$39(3) + 180(4) = 837.$$

*AMM, 1972, p. 303, Problem E2344, proposed by Jordi Dou, Barcelona, Spain.

Therefore

$$237 + 2r = 837, \text{ and } r = 300.$$

The number of blue segments, then, is $523 - 300 = 223$.

A similar problem was given in the AMM, 1971, p. 796, Problem E2251, by T. C. Brown, Simon Fraser University, and was solved by Stephen B. Maurer, Phillips Exeter Academy:

Consider a rectangular array of red dots and blue dots with an even number of rows and an even number of columns. In each row, half the dots are red and half blue; similarly for the columns. Whenever two dots of the same color are adjacent in a row or column, they are connected with a segment of their common color. Show that, altogether, the number of red segments and the number of blue segments are the same.

Solution:

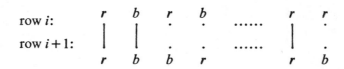

Consider two adjacent rows of dots, say row i and row $i+1$. If two dots of the same color face each other, they are paired with a connecting segment. Let us say that such dots are "matched", and that unlike pairs are "mismatched." Now, every row has half its dots red and the other half blue. Thus every row has the same number of red dots. Clearly, adjacent rows have the same number of matched red dots. Therefore they must also have the same number of mismatched red dots.

Since a mismatched red dot in row i faces a mismatched blue dot in row $i+1$, we have

the number of mismatched blue dots in row $i+1$

$=$ the number of mismatched red dots in row i

$=$ the number of mismatched red dots in row $i+1$.

In row $i+1$, then, we have the same number of mismatched red dots as mismatched blue dots. Since row $i+1$ has the same total number of red dots as blue dots, it follows that row $i+1$ must have the same number of matched red dots as matched blue dots, implying the same number of red segments as blue segments bridging the two rows. The same is true for adjacent columns, and the conclusion follows.

SWALE'S METHOD*

Given just the circumference C of a circle, it is an easy matter to construct the center and determine the radius. It is customary to draw the perpendicular bisectors of two chords in order to get the center, from which the radius is immediate. Whereas it takes a pair of arcs to determine a perpendicular bisector, the perpendicular bisectors of two abutting chords can be accomplished with three arcs (by letting the same arc serve double duty in both pairs). The total use of instruments, then, is at least 3 arcs and two applications of the straightedge. Establish the following method of constructing the radius, known as Swale's method, which employs only 2 arcs and one use of the straightedge:

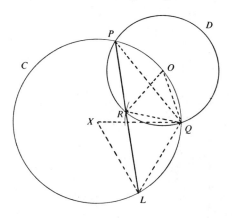

Fig. 60.

*Pi Mu Epsilon, Vol. 1, 1951, p. 146, Problem 1, proposed by Leo Moser, solved by Ding Hwang, University of California.

With any point O on the circumference C construct a circle D to intersect C at P and Q. With center Q and the same radius, cut off the point R on D inside C. Let PR meet C at L. Then QL is the radius of C (so is LR).

Solution:

Clearly all three sides of $\triangle QOR$ are the radius of D, making it equilateral and $\angle ROQ = 60°$ (Fig. 60). Thus $\angle RPQ = 30°$ at the circumference (of both circles) and the angle LXQ at the center X of C is also 60°. Therefore $\triangle XQL$ is also equilateral, and $QL = r$.

(It is not difficult to prove that triangles RQL and XOQ are congruent, from which we obtain $RL = OX = r$.)

ON $\pi(n)$

The prime numbers have long been of special interest to mathematicians. For an outstanding elementary discussion of this fascinating subject, see chapter III of W. Sierpiński's *Theory of Numbers*, pp. 110–155. One of the interesting functions concerning primes is $\pi(n)$, the number of primes not exceeding the natural number n.

Problem: Prove that $\pi(n) \geqslant \dfrac{\log n}{\log 4}$.

Solution:

The eminent Hungarian mathematician Paul Erdös found the following elementary proof of this result (see page 130 of Sierpiński's book, cited above).

Let m denote a natural number. Suppose that k^2 is the greatest square which divides m, and that $m = k^2 v$. Then v can have no repeated factors lest a square greater than k^2 divide m. (v is called the "square-free" part of m.)

Now, let a natural number n be specified and consider the natural numbers $m \leqslant n$. Let each of the numbers $m = 1, 2, \ldots, n$ be expressed in the form $m = k^2 v$, where v has no repeated factors. Since, in each case, we have $k^2 \leqslant m \leqslant n$, k must be one of the numbers $1, 2, 3, \ldots, [\sqrt{n}\,]$, where $[\sqrt{n}\,]$ denotes the greatest integer not exceeding $\sqrt{n}$. Also, since $v \leqslant m \leqslant n$, all the prime divisors of v must come from primes which are $\leqslant n$, namely $p_1, p_2, \ldots, p_{\pi(n)}$. Being the product of prime numbers from this set, v must be one of the numbers

$$p_1^{a_1} p_2^{a_2} \cdots p_{\pi(n)}^{a_{\pi(n)}},$$

where each exponent a_i is either 0 or 1 (v has no repeated factors). Since each a_i may be 0 or 1, there are $2^{\pi(n)}$ different numbers of this form. In summary, then, for each $m = k^2 v \leqslant n$, where v has no repeated factors, k must belong to the set of $[\sqrt{n}\,]$ numbers

$$X = (1, 2, 3, \ldots, [\sqrt{n}\,])$$

and v must belong to the set of $2^{\pi(n)}$ numbers

$$Y = \left(p_1^{a_1} p_2^{a_2} \cdots p_{\pi(n)}^{a_{\pi(n)}}, \qquad a_i = 0 \text{ or } 1 \right).$$

Now each of the numbers $m = 1, 2, \ldots, n$ gives rise to a k in X and a v in Y such that $m = k^2 v$. Conversely, then, by appropriately selecting k's from X and v's from Y, it is possible to construct each of the numbers $1, 2, \ldots, n$ in the form $k^2 v$. Consequently, the procedure of selecting a k from X, a v from Y, and constructing $k^2 v$, *when performed in all possible ways*, must generate a set of numbers which includes the n numbers $1, 2, \ldots, n$. But this construction generates a total of

$$[\sqrt{n}\,] \cdot 2^{\pi(n)} \quad \text{numbers.}$$

Accordingly, we have

$$[\sqrt{n}\,] \cdot 2^{\pi(n)} \geqslant n.$$

But, $\sqrt{n} \geqslant [\sqrt{n}\,]$, by definition, and therefore

$$\sqrt{n} \cdot 2^{\pi(n)} \geqslant n, \qquad 2^{\pi(n)} \geqslant \sqrt{n},$$

$$\pi(n) \cdot \log 2 \geqslant \frac{1}{2} \cdot \log n, \qquad \pi(n) \geqslant \frac{\log n}{2 \cdot \log 2} = \frac{\log n}{\log 4}.$$

In the nineteenth century, the Russian mathematician P. Tchebycheff proved the much stronger theorem:

$$\pi(n) > \frac{n}{12 \cdot \log n}.$$

(This is proved on page 149 of Sierpiński's *Theory of Numbers*.)

A short problem concerning $\pi(n)$ was given by Paul Erdös in the AMM, 1944, p. 479, Problem 4083, and solved by Whitney Scobert, University of Oregon:

If $a_1 < a_2 < \cdots < a_k \leqslant n$ is an arbitrary sequence of natural numbers such that no a_i divides the *product* of the others, prove that $k \leqslant \pi(n)$.

Solution:

Each of the a_i is $\leqslant n$ and therefore has a prime decomposition

$$a_i = p_1^{q_1} p_2^{q_2} \cdots p_{\pi(n)}^{q_{\pi(n)}}.$$

In order for a_i not to divide the product of the others, it must contain some prime $p_i \in (p_1, p_2, \ldots, p_{\pi(n)})$ to a degree which exceeds the sum of the degrees of p_i in all the other a's. Thus a_i must contain this p_i to a degree which exceeds its degree in every other individual a. Each a_i has such a p_i, and clearly no two a's can be associated in this way with the same p_i. Therefore you can't have more a's than there are primes in $(p_1, p_2, \ldots, p_{\pi(n)})$, and accordingly $k \leqslant \pi(n)$.

A CONSTANT CHORD*

Suppose two circles Q and R intersect in A and B (FIG. 61). A point P on the arc of Q which lies outside R is projected through A and B to determine chord CD of R. Prove that no matter where P is chosen on its arc, the length of the chord CD is always the same.

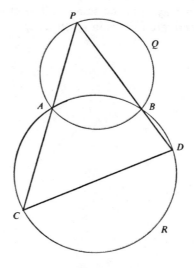

FIG. 61.

*AMM, 1933, p. 265, "On Two Intersecting Spheres" by N. A. Court, University of Oklahoma.

Solution:

Let P and P' denote positions of P corresponding to chords CD and $C'D'$ (FIG. 62). We have
$$\angle PAP' = \angle PBP',$$
$$\angle PAP' = \angle CAC',$$
$$\angle PBP' = \angle DBD',$$
giving $\angle CAC' = \angle DBD'$. This means arc $CC' =$ arc DD'. Adding arc CD' to each gives
$$\text{arc } C'D' = \text{arc } CD,$$
from which we have
$$C'D' = CD.$$

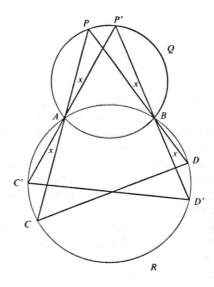

FIG. 62.

THE NUMBER OF INNER DIAGONALS*

A simple polygon is one which does not intersect itself. However, a simple n-gon may be far from convex and have many diagonals lying outside or partly outside of it. Prove that, nevertheless, every simple n-gon has at least $n-3$ diagonals which lie completely inside it.

Solution:

Clearly the claim is valid for quadrilaterals (FIG. 63). Suppose the claim is valid for k-gons, $k=4,5,\ldots,n$. Consider the simple $(n+1)$-gon P. Now it is impossible for a polygon to have none of its diagonals completely inside it. For a proof of this result see Ross Honsberger: *Ingenuity in Mathematics*, Vol. 23, New Mathematical Library, Mathematical Association of America, page 35. Let d denote a diagonal lying wholly inside P. Suppose d divides P into an r_1-gon and an r_2-gon. The induction hypothesis, then, gives at least r_1-3 and r_2-3 interior diagonals of P in these two subpolygons. Counting d, itseif, there must be at least r_1+r_2-5 interior diagonals in P.

Now, the total number of sides in the two subpolygons is r_1+r_2. This includes all $n+1$ sides of P and it counts the diagonal d twice as well. Thus we have $r_1+r_2=n+3$. Therefore P has at least

$$r_1+r_2-5=n+3-5=n-2=(n+1)-3$$

interior diagonals, and the conclusion follows by induction.

Since we can exhibit, for each natural number n, an n-gon which has *exactly* $n-3$ interior diagonals, we see that $n-3$ is the most such diagonals we can always count on. Let tangents to a

---

*AMM, 1970, p. 1111, Problem E2214, proposed by Murray Klamkin, Ford Scientific Laboratory, and B. Ross Taylor, York High School.

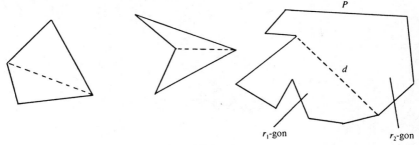

FIG. 63.

circle C at A and B meet at P (FIG. 64). Choosing $N_1, N_2, \ldots, N_{n-3}$ along the arc AB which is nearer P, we obtain an n-gon $PAN_1N \ldots N_{n-3}B$ which has precisely $n-3$ interior diagonals PN_i, all emanating from P.

R. B. Eggleton of Australia has established that a simple n-gon has exactly $n-3$ inner diagonals if and only if no two of its inner diagonals intersect.

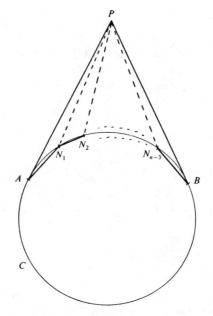

FIG. 64.

LOADING THE DICE*

Prove that it is impossible to load a pair of dice so that every sum 2, 3,..., 12 is equally likely. As customary, assume that the dice are distinguishable (e.g., a 2 on the first die with a 4 on the second is different from a 4 on the first and a 2 on the second, even though the same total 6 is obtained).

Solution:

Let p_i denote the probability of i coming up on the first die and q_i the probability of i on the second die. The probability of getting the sum 2, then, is $p_1 q_1$, while that of getting the sum 12 is $p_6 q_6$. If all eleven of the probabilities are the same, each would be $1/11$. The probability of getting 7 is

$$\frac{1}{11} = p_1 q_6 + p_2 q_5 + \cdots + p_6 q_1$$

$$\geq p_1 q_6 + p_6 q_1 = p_1 q_6 \left(\frac{q_1}{q_1}\right) + p_6 q_1 \left(\frac{q_6}{q_6}\right)$$

$$= p_1 q_1 \left(\frac{q_6}{q_1}\right) + p_6 q_6 \left(\frac{q_1}{q_6}\right) = \frac{1}{11}\left(\frac{q_6}{q_1}\right) + \frac{1}{11}\left(\frac{q_1}{q_6}\right)$$

$$= \frac{1}{11}\left(\frac{q_6}{q_1} + \frac{q_1}{q_6}\right).$$

*AMM, 1951, p. 191, Problem E925, proposed by J. B. Kelly, University of Wisconsin, solved by Leo Moser and J. H. Wahab, University of North Carolina.

Therefore we obtain

$$\frac{q_6}{q_1} + \frac{q_1}{q_6} \leqslant 1.$$

But the sum of a positive real number and its reciprocal $x + 1/x$ is always at least 2. Therefore the proposed loading is impossible.

A CURIOUS SEQUENCE*

Suppose we pass over the natural numbers $1, 2, 3, \ldots$, and pick out a sequence U by taking

the first odd number,	(namely 1)
the next two even numbers,	(2 and 4)
the next three odd numbers,	(5, 7, 9)
the next four even numbers,	(10, 12, 14, 16)
the next five odd numbers,	(17, 19, 21, 23, 25)
and so on.	

$$U: \quad 1, 2, 4, 5, 7, 9, 10, 12, 14, 16, 17, 19, \ldots .$$

Prove that the nth term, u_n, is given by the formula

$$u_n = 2n - \left[\frac{1 + \sqrt{8n - 7}}{2} \right],$$

where $[x]$ denotes the greatest integer not exceeding x.

Solution:

Let us compare U with the sequence E of even numbers, and let the sequence of differences between corresponding terms be $D \equiv \{d_n\}$. We shall see that D consists of one 1, two 2's, three 3's,...,

$E \equiv \{2n\}$	:	2,	4, 6,	8, 10, 12,	14, 16, 18, 20,	22,...,
$U \equiv \{u_n\}$	:	1,	2, 4,	5, 7, 9,	10, 12, 14, 16,	17,...,
$D \equiv \{d_n\}$	:	1	2, 2,	3, 3, 3,	4, 4, 4, 4,	5,...,

---

*AMM, 1960, p. 380, Problem E1382, proposed by Ian Connell, University of Manitoba, solved by Andrew Korsak, University of Toronto.

n n's,.... The terms of U are defined in groups of one term, two terms, three terms, etc. Within a group, the terms, whether odd or even, increase by 2's, as do the even numbers. Thus the difference sequence D remains constant throughout a group. In passing to the next group, however, the even numbers continue to increase by 2 while the given sequence U, in shifting from odd to even, or vice-versa, increases by only 1. Accordingly, the difference which persists through the next group grows by 1, and the terms of D are as claimed above. By determining a formula for d_n, we obtain the required formula for u_n from the relation

$$u_n = 2n - d_n.$$

For a specified n, then, let us try to determine d_n. As we have observed, the terms of D occur in groups of repeated values:

$$(1), (2,2), (3,3,3), \ldots, \underbrace{(k-1, k-1, \ldots)}_{k-1 \text{ times}}, \underbrace{(k, k, \ldots)}_{k \text{ times}}, \ldots.$$

We need to find out the group into which our particular d_n falls. If it falls into the kth group, then its value is k. We observe that, preceding the kth group, we have

one 1, two 2's, three 3's,$\ldots, k-1$ $(k-1)$'s,

a total of

$$1 + 2 + 3 + \cdots + (k-1) = \frac{(k-1)k}{2} \text{ terms.}$$

If d_n falls into the kth group, then, being the nth term in D, we must have

$$\frac{(k-1)k}{2} + 1 \leqslant n < \frac{[(k+1)-1](k+1)}{2} + 1.$$

For example, if d_n is in the 10th group, then

$$\frac{(10-1)10}{2} + 1 \leqslant n < \frac{(11-1)11}{2} + 1.$$

Of course, in this case, we also have $(9-1)9/2 + 1 \leqslant n$, $(8-1)8/2 + 1 \leqslant n, \ldots, (1-1)1/2 + 1 \leqslant n$. Of all the integers $(m-1)m/2 + 1$, it is $m = k$ which provides the *greatest* value $\leqslant n$. Thus we have

$k = d_n$ to be the greatest integer m satisfying $(m-1)m/2 + 1 \leqslant n$, or

$$m^2 - m + 2(1 - n) \leqslant 0.$$

Now, $m^2 - m + 2(1 - n)$ is nonnegative for all values of m lying in the closed interval between the roots of the corresponding equation $m^2 - m + 2(1 - n) = 0$, which are

$$m = \frac{1 \pm \sqrt{1 - 8(1 - n)}}{2} = \frac{1 \pm \sqrt{8n - 7}}{2} \qquad \text{(FIG. 65)}.$$

Thus k is the greatest integer in the range

$$\frac{1 - \sqrt{8n - 7}}{2} \leqslant m \leqslant \frac{1 + \sqrt{8n - 7}}{2}.$$

Since k is an integer, we have

$$d_n = k = \left[\frac{1 + \sqrt{8n - 7}}{2} \right],$$

and the required formula is

$$u_n = 2n - \left[\frac{1 + \sqrt{8n - 7}}{2} \right].$$

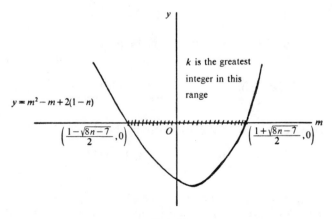

FIG. 65.

Nathan Mendelsohn (University of Manitoba) directed attention to the complementary sequence $V \equiv \{v_n\}$ of natural numbers which are not in U:

$$V \equiv \{v_n\} : 3, 6, 8, 11, 13, 15, 18, 20, 22, 24, 27, \ldots.$$

He pointed out that the terms of V also go together in groups of one odd term, two even terms, three odd, etc., and that a formula for v_n is obtained, not by subtracting d_n from $2n$, as in the case of u_n, but by adding:

$$v_n = 2n + d_n = 2n + \left[\frac{1 + \sqrt{8n - 7}}{2} \right].$$

This follows immediately from the fact that $u_n + v_n = 4n$. It is not very difficult to prove this result, and it is left as an exercise.

LONG STRINGS OF CONSECUTIVE
NATURAL NUMBERS*

Although there exists an infinity of prime numbers, the gaps in the natural sequence between consecutive primes grow arbitrarily large. This is easily seen to be so by observing that, for all natural numbers n, the numbers

$$(n+1)! + 2, (n+1)! + 3, \ldots, (n+1)! + (n+1)$$

constitute a set of n consecutive composite numbers.

Prove that there also exist arbitrarily long stretches of consecutive natural numbers each of which is divisible by a perfect square exceeding 1.

Solution:

We shall show by induction that, for all natural numbers n, there exists a set of n consecutive natural numbers each of which is divisible by a perfect square > 1.

(i) For $n = 1$, any square > 1 satisfies our requirement.

(ii) Suppose, for $n \geqslant 1$, the n consecutive natural numbers

$$a_1, a_2, a_3, \ldots, a_n$$

are each divisible by a perfect square > 1. Now we look for $n+1$ consecutive numbers with the same property.

Let s_i denote a perfect square > 1 which divides a_i, $i = 1, 2, \ldots, n$, and let L denote the product of the s_i. Because the a_i are consecu-

---

*MM, 1952, p. 221, Problem 106, proposed by E. P. Starke, Rutgers University, solved by S. B. Akers, Jr., U. S. Coast Guard Headquarters, Washington D. C.

tive, we have $a_2 = a_1 + 1$, and so on. In keeping with this notation, let $a_n + 1$ be denoted a_{n+1}, giving $a_1, a_2, \ldots, a_n, a_{n+1}$ to be a string of $n+1$ consecutive numbers. Let $a_{n+1}(L+2)L$ be denoted by A. Containing the factor L, A is divisible by each s_i. Now consider the $n+1$ consecutive natural numbers

$$A + a_1, A + a_2, \ldots, A + a_{n+1}.$$

For $i = 1, 2, \ldots, n$ we have s_i dividing both A and a_i, showing that the first n of these numbers are each divisible by a square > 1. And the last number is

$$A + a_{n+1} = a_{n+1}(L+2)L + a_{n+1}$$
$$= a_{n+1}(L^2 + 2L + 1) = a_{n+1}(L+1)^2.$$

Since $s_i > 1$, then $L > 1$, and $(L+1)^2$ is a square > 1. Thus all $n+1$ of the numbers are divisible by a square > 1 and the conclusion follows by induction.

It can be shown in the same way that, for all natural numbers n, there exist n consecutive natural numbers each of which is divisible by a perfect mth power > 1, where $m > 2$. We need only take the s_i to be mth powers > 1 and alter A to

$$A = a_{n+1}\left[(L+1)^m - 1\right].$$

Observe that our argument not only establishes the existence of any desired sequence of the types discussed but shows how to construct the sequence by iteration. Since the only property of L that is needed is that it be divisible by each s_i, instead of taking L to be the product of all the s_i we may use their least common multiple. This may be a lesser number and reduce the calculations.

It strikes me as truly remarkable that somewhere along the sequence of natural numbers there are a trillion numbers (or however many you like) in a row each of which is divisible by a perfect trillionth power > 1, and that, given enough time, I could actually produce such a string for you!

A MINIMAL INSCRIBED
QUADRILATERAL*

$ABCD$ is a cyclic quadrilateral whose diagonals meet at X. P, Q, R, S are the feet of the perpendiculars from X upon the sides of $ABCD$. Prove that, of all quadrilaterals having a point on each side of $ABCD$, $PQRS$ has minimum perimeter.

Solution:

We begin by proving that PS and PQ make equal angles with AB (FIG. 66):

$$\angle APS = \angle BPQ.$$

We do this by showing that they have equal complements:

$$\angle SPX = \angle QPX.$$

The right angles clearly make $PBQX$ cyclic, giving $\angle QPX = \angle QBX$. Similarly, $APXS$ cyclic and $\angle SPX = \angle SAX$. However, in the given circle, $\angle CBD = \angle CAD$. Hence $\angle QPX = \angle SPX$, and the conclusion follows. Similarly, at every vertex of $PQRS$, the sides make equal angles with the corresponding side of $ABCD$.

The consequence of this is that, reflecting $PQRS$ in a side of $ABCD$, the image of either of the sides which meet on the mirror is just the extension of the other such side (FIG. 67).

Accordingly, we can straighten out the perimeter of $PQRS$ in a series of three reflections I, II, and III, as illustrated (FIG. 68). We

*AMM, 1926, p. 161, Problem 2728, proposed by G. Y. Sosnow, Newark, New Jersey, solved by Michael Goldberg.

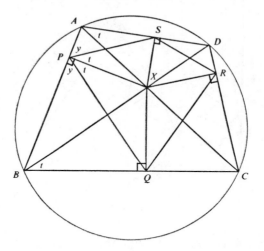

FIG. 66.

(Since APB is straight,
$2y + \angle SPQ = 180°$. But
the reflection makes
$\angle BPQ_1 = \angle BPQ = y$.
Thus
$\angle Q_1PS = 2y + \angle SPQ = 180°$.)

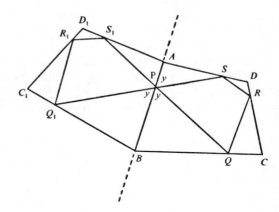

FIG. 67.

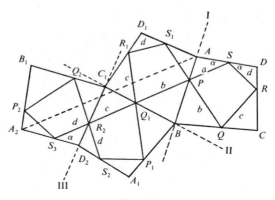

FIG. 68.

have here, of course, just four abutting congruent pictures of
$PQRS$ inside $ABCD$. Thus $A_2S_3 = AS$. But the alternate angles
$D_2S_3R_2$ and ASP are equal (each equals $\angle DSR$), implying that
they are parallel as well. Therefore A_2ASS_3 is a parallelogram,
making the straightened out perimeter SS_3 of $PQRS$ equal to AA_2.

Now these same reflections unfold the perimeter of any quadrilateral having a vertex on each side of $ABCD$. If Y is the vertex on
AD and Y_3 its image on A_2D_2, then, because A_2Y_3 and AY are
equal and parallel, we see again that A_2AYY_3 is a parallelogram
(FIG. 69). Thus $YY_3 = AA_2$. However, the unfolded perimeter ex-

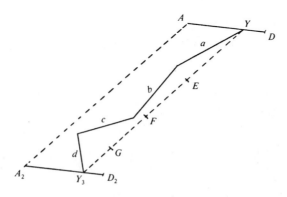

FIG. 69.

tends between Y and Y_3. If it forms a crooked arc, then it is longer than YY_3. At its shortest, it could only coincide with YY_3, which equals AA_2, the perimeter of $PQRS$. Thus $PQRS$ has minimum perimeter.

Observe that if E, F, and G are the points on YY_3 where it crosses AB, BC_1, and C_1D_2, respectively, then E, F, and G correspond to points on the other three sides of $ABCD$, determining an inscribed quadrilateral T. Unfolding T simply yields YY_3 as its straightened perimeter. Thus from any point Y on AD there exists a quadrilateral inscribed in $ABCD$ which has minimum perimeter.

TRIANGULAR NUMBERS

The numbers of disks in the triangular arrays below determine the sequence of triangular numbers (FIG. 70). They begin $1, 3, 6, 10, 15, 21, 28, 36, 45, \ldots$, the nth one being

$$t_n = 1 + 2 + 3 + \cdots + n = \frac{n(n+1)}{2}.$$

Since $t_n = t_{n-1} + n$, the first couple of dozen terms are easily determined mentally:

1, (add 2) 3, (add 3) 6, (add 4) 10, (add 5) 15, (add 6)
$21, 28, 36, 45, 55, 66, 78, 91, 105, 120, 136, 153, \ldots$.

In this section we consider 7 little problems about triangular numbers.

(i) (*MM*, 1935–36, *p.* 313, *Problem* 115, *proposed and solved by G. W. Wishard, Norwood, Ohio.*)

Prove that every odd square in the octonary system (scale of 8) ends in 1, and if this 1 be cut off, the remaining part is always a triangular number.

Solution:

Observing that one of the consecutive numbers n and $n+1$ is even, we have

$$(2n+1)^2 = 4n^2 + 4n + 1 = 4n(n+1) + 1 = 8k + 1,$$

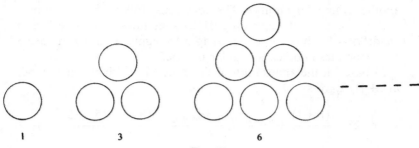

1 3 6

FIG. 70.

for some integer k. In the octonary system, then, an odd square ends in 1.

Cutting off the last digit 1 from a number m in the scale of 8 leaves the number $(m-1)/8$. Thus the proposed deletion yields

$$\frac{(2n+1)^2-1}{8} = \frac{4n(n+1)}{8} = \frac{n(n+1)}{2} = t_n,$$

the nth triangular number.

(ii) (*AMM*, 1932, *p*. 179, *Problem* 3480, *proposed by G. W. Wishard, Norwood, Ohio, solved by Helen A. Merrill, Wellesley College.*)

Prove that, in base 9, every number consisting entirely of 1's is a triangular number:

$$1, \ 11, \ 111, \ 1111, \ldots.$$

Solution:

Clearly the first number, 1, is triangular. Now, if 1 is appended at the end of a number k expressed in base 9, the result is the number $9k+1$. If k happens to be a triangular number, $n(n+1)/2$, then this procedure produces

$$9 \cdot \frac{n(n+1)}{2} + 1 = \frac{9n^2+9n+2}{2} = \frac{(3n+1)(3n+2)}{2},$$

another triangular number. The conclusion follows by induction.

We observe that appending 01 to a number k in base 3 also yields $9k + 1$. Thus, attaching 01 to a triangular number in base 3 also generates another triangular number.

In base 25, the annexation of the digit 3 to a triangular number $n(n+1)/2$ gives

$$25 \cdot \frac{n(n+1)}{2} + 3 = \frac{25n^2 + 25n + 6}{2} = \frac{(5n+2)(5n+3)}{2},$$

a triangular number. Hence, appending 3 in base 25, or the equivalent 03 in base 5, to a triangular number produces another triangular number.

In general, if a triangular number in base $(2k+1)^2$ has the digit $\frac{1}{2}k(k+1)$ annexed; or, in base $2k+1$, $k \leqslant 3$, has the digits 0 and $\frac{1}{2}k(k+1)$ annexed; or, in base $2k+1$, $k > 3$, has the two digits representing the number $\frac{1}{2}k(k+1)$ annexed; then the resulting number is triangular.

(iii) (*AMM*, 1963, p. 211, *Problem E*1516, *proposed by R. J. Oberg, University of California, solved by J. E. Yeager, Temple University*.)

Let $N = 0.1360518\ldots$ be formed from the last digits of the triangular numbers. Is N rational or irrational?

Solution:

The triangular numbers proceed

$$1, 3, 6, 10, 15, 21, 28, 36, 45, 55, 66, 78, 91, 105, 120, 136, 153, 171,$$
$$190, 210, 231, 253, 276, 300, 325, 351, \ldots .$$

Thus we see that N proceeds

$$N = .13605186556815063100136051\ldots$$

suggesting a period of 20 digits. Continuing to denote the nth triangular number by t_n, we have in general that

$$t_{n+20} - t_n = \frac{(n+20)(n+21)}{2} - \frac{n(n+1)}{2}$$

$$= \frac{1}{2}(n^2 + 41n + 420 - n^2 - n)$$

$$= 10(2n+21), \quad \text{which ends in 0.}$$

Thus t_{n+20} and t_n must have the same last digit, confirming the suspicion that N is periodic. Hence N is rational.

We observe that the period of N is almost a palindrome (the same backwards as forwards). In fact,

$$\frac{N}{10} = .01360518655681506310013605\ldots$$

has a palindromic period.

We note that it can be proved that a rational number N is obtained from the last k digits of the triangular numbers for $k = 1, 2, 3, \ldots$. The triangular numbers are the partial sums of the arithmetic progression $1, 2, 3, \ldots$. It is not difficult to show that the number N, constructed from the last k digits of the partial sums of any arithmetic progression of natural numbers, is a rational number.

(iv) (*AMM*, 1962, *p.* 168, *Problem E*1473, *proposed by J. L. Pietenpol, Columbia University, solved by A. V. Sylwester, U. S. Naval Ordnance Laboratory, Corona, California.*)

We can see that the triangular numbers 1 and 36 are also perfect squares. Prove that an infinity of the triangular numbers are also perfect squares.

Solution:

This follows the penetrating observation that $t_{4n(n+1)}$ is a perfect square whenever t_n is:

$$\text{if } t_n = \frac{n(n+1)}{2} = k^2, \quad \text{then} \quad 4n(n+1) = 8k^2,$$

and

$$t_{4n(n+1)} = t_{8k^2} = \frac{8k^2(8k^2+1)}{2} = 4k^2(8k^2+1)$$
$$= 4k^2[4n(n+1)+1] = 4k^2(4n^2+4n+1)$$
$$= 4k^2(2n+1)^2, \quad \text{a square.}$$

(v) (*AMM, 1933, p. 362, Problem E21, proposed by V. F. Ivanoff, San Francisco, solved by L. S. Johnston, University of Detroit.*)

Prove that the difference between the squares of two consecutive triangular numbers is always a perfect cube:

$$\{t_n\}: \qquad 1, \ 3, \ 6, \ 10, \ 15, \ 21, \dots,$$
$$\{t_n^2\}: \qquad 1, \ 9, \ 36, \ 100, \ 225, \ 441, \dots,$$
$$\{\text{differences}\}: \qquad 8, \ 27, \ 64, \ 125, \ 216, \dots.$$

Solution:

It is well known that the sum of the cubes of the first n natural numbers is

$$1^3 + 2^3 + \cdots + n^3 = \left[\frac{n(n+1)}{2}\right]^2 = t_n^2.$$

Thus $t_{n+1}^2 - t_n^2 = (n+1)^3$, as required.

(vi) (*Pi Mu Epsilon, Vol. 2, 1954–59, p. 378, Problem 92, proposed by Leon Bankoff, Los Angeles, solved by Thomas Porsching, Carnegie Institute of Technology.*)

Prove that the sum of the reciprocals of the triangular numbers is 2:

$$\frac{1}{1} + \frac{1}{3} + \frac{1}{6} + \frac{1}{10} + \cdots = 2.$$

Solution:

A lovely geometric solution is obtained from the two hyperbolas

$$y_1(x) = \frac{1}{x} \quad \text{and} \quad y_2(x) = \frac{1}{x-1} \qquad \text{(FIG. 71)}.$$

For x an integer $n \geqslant 2$, the difference a_n between the ordinates to these curves is

$$a_n = \frac{1}{n-1} - \frac{1}{n} = \frac{1}{n(n-1)} = \frac{1}{2} \cdot \frac{2}{n(n-1)} = \frac{1}{2}\left(\frac{1}{t_{n-1}}\right).$$

Thus $1/t_{n-1} = 2a_n$, and the sum in question is

$$\frac{1}{t_1} + \frac{1}{t_2} + \cdots = 2(a_2 + a_3 + \cdots).$$

Now the value of $y_2(x)$ at $x = n$ is the same as the value of $y_1(x)$ at $x = n - 1$. Hence the projections of the a_n onto the y-axis fit together end-to-end to build up a continuous segment from $(0, 1)$ down toward the origin. Since both hyperbolas are asymptotic to the positive x-axis, these projections completely fill the unit seg-

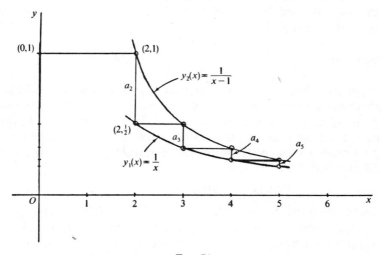

FIG. 71.

ment to the origin, implying that

$$a_2 + a_3 + a_4 + \cdots = 1,$$

from which the desired conclusion follows immediately.

In addition to this geometric solution there is the following elegant algebraic approach:

$$
\begin{aligned}
\frac{1}{1} + \frac{1}{3} + \frac{1}{6} + \frac{1}{10} + \cdots &= \frac{2}{1\cdot2} + \frac{2}{2\cdot3} + \frac{2}{3\cdot4} + \frac{2}{4\cdot5} + \cdots \\
&= 2\left(\frac{1}{1\cdot2} + \frac{1}{2\cdot3} + \frac{1}{3\cdot4} + \frac{1}{4\cdot5} + \cdots\right) \\
&= 2\left[\left(1 - \frac{1}{2}\right) + \left(\frac{1}{2} - \frac{1}{3}\right) + \left(\frac{1}{3} - \frac{1}{4}\right) + \cdots\right] \\
&= 2.
\end{aligned}
$$

(vii) (*AMM*, 1968, *p.* 410, *Problem E*1943, *proposed by J. M. Khatri, Baroda, India, solved by Bernard Jacobson, Franklin and Marshall College.*)

Our last problem inquires whether it is possible to pick out an infinite sequence of triangular numbers with the property that the sum of any number of terms from the beginning of the sequence is, itself, a triangular number.

Solution:

Suppose that we have built up part of such a sequence and that, in addition to the required property, our start enjoys the special property that the sum up to any term t_k in this part is t_{k+1}. In this case, the annexation of the term $t_{t_{k+1}-1}$ extends the sequence while preserving its essential properties. Because $t_n + (n+1) = t_{n+1}$, it follows that, for $n = t_{k+1} - 1$,

$$\text{the new sum} = t_{t_{k+1}-1} + t_{k+1} = t_{t_{k+1}}.$$

Such a sequence is seeded by $t_3, t_5, t_{20}, t_{230}$, which has partial sums $(t_3), t_6, t_{21}$, and t_{231}. In fact, beginning with any triangular number beyond the second, this rule of annexation generates an infinite sequence with the required property.

We note in closing that it is also possible to select an infinite sequence of triangular numbers with the property that all of its partial sums are perfect squares. An example is given by $t_1, t_2, t_6, t_{18}, \ldots, t_{2 \cdot 3^k}, \ldots$, for which

$$t_1 + t_2 + t_6 + \cdots + t_{2 \cdot 3^k} = \left(\frac{3^{k+1} + 1}{2} \right)^2.$$

ON A REGULAR n-GON*

Let $A_1A_2 \cdots A_n$ denote a regular n-gon. Prove that, no matter where a point O is chosen inside the polygon, at least one of the angles A_iOA_j is as close to a straight angle as $(1/n)$th of a straight angle:

$$\pi - \frac{\pi}{n} \leqslant \angle A_iOA_j \leqslant \pi.$$

Solution:

Let A_1 denote a vertex for which the distance OA_i is a minimum, that is, $OA_1 \leqslant OA_i$ for $i = 2, 3, \ldots, n$ (FIG. 72). Join A_1 to each of the other vertices. Now if O were to lie on one of the diagonals A_1A_i, we see immediately that $\angle A_1OA_i = \pi$. Suppose, then, that O lies between the consecutive diagonals A_1A_i and A_1A_{i+1}. Let the angles in triangles A_1OA_i and A_1OA_{i+1} be denoted x, y, z, t, m, n as shown. Because $A_1O \leqslant OA_i$, we have $z \leqslant m$. Similarly, $A_1O \leqslant OA_{i+1}$ gives $t \leqslant n$. Hence $z + t \leqslant m + n$. But a regular n-gon is cyclic. Let C denote the center of the circumcircle of the polygon. Then the arc A_iA_{i+1}, joining consecutive vertices, subtends at the center C an angle of $2\pi/n$. Consequently, at the circumference, the angle is one-half as big, and we have $m + n = \pi/n$. Thus $z + t \leqslant \pi/n$.

---

*AMM, 1947, p. 117, Problem 4086, proposed by Paul Erdös, University of Michigan, solved by C. R. Phelps, Rutgers University.

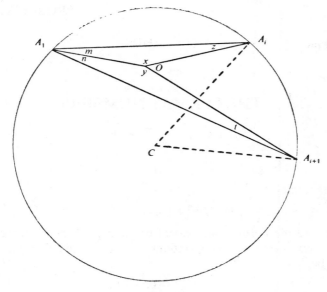

FIG. 72.

However, in the two triangles, the six angles sum to 2π:

$$(m+n)+(z+t)+x+y=2\pi,$$
$$\frac{\pi}{n}+(z+t)+x+y=2\pi.$$

Since $z+t \leqslant \pi/n$, we have that

$$\frac{\pi}{n}+\frac{\pi}{n}+x+y \geqslant 2\pi,$$

or

$$x+y \geqslant 2\left(\pi-\frac{\pi}{n}\right).$$

Thus, not both of x and y can be less than $\pi-(\pi/n)$ (lest their sum be too small), and the conclusion follows.

THE FERMAT NUMBERS

The numbers

$$F_n = 2^{(2^n)} + 1, n = 0, 1, 2, \ldots,$$

are called the Fermat numbers after the great French mathematician Pierre de Fermat (1601–1665). The sequence of Fermat numbers begins

$$3, 5, 17, 257, 65537, \ldots.$$

They satisfy the recursive relation $F_n = F_0 F_1 \cdots F_{n-1} + 2$. This is easily established by induction. However, the following neat way is given in AMM, 1935, p. 569, Problem E152, proposed by J. Rosenbaum, Hartford Federal College, Connecticut, and solved by Daniel Finkel, Brooklyn, New York.

Observe that $2^{(2^0)} - 1$ is simply the number 1. Then

$$
\begin{aligned}
1 \cdot F_0 F_1 \cdots F_{n-1} &= [2^{(2^0)} - 1][2^{(2^0)} + 1][2^{(2^1)} + 1] \cdots [2^{(2^{n-1})} + 1] \\
&= [2^{(2^1)} - 1][2^{(2^1)} + 1][2^{(2^2)} + 1] \cdots [2^{(2^{n-1})} + 1] \\
&= [2^{(2^2)} - 1][2^{(2^2)} + 1] \cdots [2^{(2^{n-1})} + 1] \quad \cdots \\
&= [2^{(2^{n-1})} - 1][2^{(2^{n-1})} + 1] \\
&= 2^{(2^n)} - 1 = F_n - 2.
\end{aligned}
$$

From this relation it is a simple matter to show that each pair of Fermat numbers are relatively prime. If $m < n$, then

$$F_n = F_0 F_1 \cdots F_m \cdots F_{n-1} + 2.$$

Consequently, any common divisor of F_m and F_n must also divide

2. Such a common divisor, then, must be either 1 or 2. But it cannot be 2 since all the Fermat numbers are odd. Hence it must be 1, making F_m and F_n relatively prime.

Thus, since each $F_n > 1$, each Fermat number has a prime divisor which does not divide any other Fermat number. Since there is an infinity of Fermat numbers, we have another proof of the infinity of the primes.

Our result also provides an immediate solution to the following problem:

Show that $2^{(2^n)} - 1$ is divisible by at least n different prime divisors.

(*AMM*, 1968, p. 1016, *Problem E*2014, *proposed by Erwin Just and Norman Schaumberger, Bronx Community College, New York City*.)

Solution:

$$2^{(2^n)} - 1 = \left[2^{(2^n)} + 1 \right] - 2 = F_n - 2$$
$$= F_0 F_1 \cdots F_{n-1}.$$

Since this is the product of n different Fermat numbers, it has at least n different prime divisors since the Fermat numbers are relatively prime.

While we are on the subject of the Fermat numbers, let us establish the easy results that, among the Fermat numbers, there are no squares, no cubes and, except for $F_0 = 3$, no triangular numbers.

(i) F_n *is never a square*.

Clearly

$$(F_n - 1)^2 = \left[2^{(2^n)} \right]^2 = 2^{(2^{n+1})} = F_{n+1} - 1.$$

Therefore

$$F_{n+1} = 1 + (F_n - 1)^2.$$

Consequently, if $F_n \equiv 2$ (mod 3), then so is $F_{n+1} \equiv 2$ (mod 3). But $F_1 = 5 \equiv 2$ (mod 3), implying that, for $n > 0$, all $F_n \equiv 2$ (mod 3). However, no square is ever congruent to 2 (mod 3) (for $n \equiv 0$, 1, or -1 (mod 3), we have $n^2 \equiv 0$, 1, or 1 (mod 3)). Since $F_0 = 3$ is not a square, it follows that no F_n is a square.

(ii) F_n *is never a cube.*

It is easily seen that a cube is congruent to 0, 1, or -1 (mod 7):

$$n \equiv 0, 1, 2, 3, 4, 5, 6,$$
$$n^2 \equiv 0, 1, 4, 2, 2, 4, 1,$$
$$n^3 \equiv 0, 1, 1, -1, 1, -1, -1.$$

For the Fermat numbers, we have $F_0 = 3$ and $F_1 = 5$. From

$$F_{n+1} = 1 + (F_n - 1)^2$$

we see that,

if $F_n \equiv 3$ (mod 7), then $F_{n+1} \equiv 5$ (mod 7),

and

if $F_n \equiv 5$ (mod 7), then $F_{n+1} \equiv 3$ (mod 7).

Consequently, modulo 7 the Fermat numbers alternate between the residues 3 and 5, and are never congruent to 0, 1, or -1. Thus there are no cubes among the Fermat numbers.

(iii) *No $F_n > 3$ is a triangular number.*

The nth triangular number is $t_n = n(n+1)/2$, giving $2t_n = n(n+1)$. Now, $n \equiv 0$, 1, or 2 (mod 3). If $n \equiv 0$ or 2 (mod 3), then one of n and $n+1$ is divisible by 3, making $t_n \equiv 0$ (mod 3). On the other hand, if $n \equiv 1$ (mod 3), we have $2t_n \equiv n(n+1) \equiv 2$ (mod 3). Now, unless $t_n \equiv 1$, this would not be so. In all cases, then, we have $t_n \equiv 0$ or 1 (mod 3), which, as we have seen above, is never true of a Fermat number > 3. The conclusion follows.

A PERFECT 4TH POWER*

Prove that the product of 8 consecutive natural numbers is never a perfect fourth power.

Solution:

Let x denote the least of 8 consecutive natural numbers. Then their product P may be written

$$P = \big[x(x+7) \big] \big[(x+1)(x+6) \big] \big[(x+2)(x+5) \big] \big[(x+3)(x+4) \big]$$
$$= (x^2+7x)(x^2+7x+6)(x^2+7x+10)(x^2+7x+12).$$

Letting $x^2+7x+6=a$, we have

$$P = (a-6)(a)(a+4)(a+6)$$
$$= (a^2-36)(a^2+4a) = a^4 + 4a^3 - 36a^2 - 144a$$
$$= a^4 + 4a(a^2-9a-36) = a^4 + 4a(a+3)(a-12).$$

Since $a = x^2+7x+6$ and $x \geqslant 1$, we have $a \geqslant 14$ and $a-12$ is positive. Hence

$$P > a^4.$$

However, $P = a^4 + 4a^3 - 36a^2 - 144a$ reveals that P is less than $(a+1)^4 = a^4 + 4a^3 + 6a^2 + 4a + 1$. Hence

$$a^4 < P < (a+1)^4,$$

showing that P always falls *between consecutive* fourth powers and never coincides with one.

*AMM, 1936, p. 310, Problem 3703, proposed by Victor Thébault, Le Mans, France, solved by the Mathematics Club of the New Jersey College for Women, New Brunswick, New Jersey.

AN INEQUALITY OF RECIPROCALS*

For n a natural number > 1, show that

$$\frac{1}{n} + \frac{1}{n+1} + \frac{1}{n+2} + \cdots + \frac{1}{n^2} > 1.$$

Solution:

$$\frac{1}{n} + \frac{1}{n+1} + \frac{1}{n+2} + \cdots + \frac{1}{n^2} > \frac{1}{n} + \left(\frac{1}{n^2} + \frac{1}{n^2} + \cdots + \frac{1}{n^2} \right)$$

$$= \frac{1}{n} + \frac{1}{n^2}(n^2 - n)$$

$$= \frac{1}{n} + 1 - \frac{1}{n}$$

$$= 1.$$

*MM, 1960, p. 244, Problem Q279, proposed by Barney Bissinger.

PACKING SQUARES*

Because the harmonic series diverges, a set of squares of sides $1, 1/2, 1/3, \ldots, 1/n, \ldots$, placed side-by-side, extend indefinitely far along their common base-line L (FIG. 73). Prove, however, that all the squares beyond the first can be packed into the first one without overlapping.

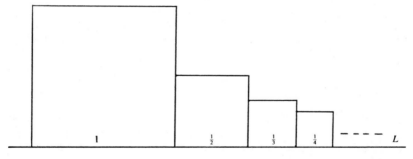

FIG. 73.

Solution:

Along L, divide the squares into groups at the denominators which are powers of 2: $(1/2, 1/3)$, $(1/4, 1/5, 1/6, 1/7), \ldots$. The sum of the sides of the squares in the nth group is

$$\frac{1}{2^n} + \frac{1}{2^n+1} + \cdots + \frac{1}{2^{n+1}-1} < \underbrace{\frac{1}{2^n} + \frac{1}{2^n} + \cdots + \frac{1}{2^n}}_{2^n \text{ times}} = 1.$$

*Pi Mu Epsilon, 1959–64, Vol. 3, p. 473, Problem 137, proposed by Leo Moser, solved by Michael Goldberg.

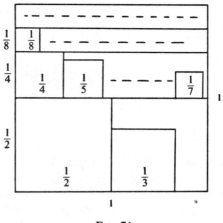

FIG. 74.

Therefore the squares of the nth group fit into a rectangle of height $1/2^n$ and width 1. Piling the rectangles containing the groups of squares on top of each other, a rectangular stack is obtained which has width 1 and total height

$$\frac{1}{2} + \frac{1}{2^2} + \frac{1}{2^3} + \cdots = 1, \quad \text{(FIG. 74)}.$$

Thus arranged in a unit square, then, the required packing is accomplished.

Our main interest in this section is the following problem:

Consider a set of squares, finite or infinite in number, whose total area is 1. Prove that, whatever their sides, they can all be packed into a square S of side $\sqrt{2}$.

(*AMM*, 1969, *p*. 88, *Problem E* 2041, *proposed and solved by D. J. Newman, Yeshiva University*.)

Solution:

Since a square of area $1/2$ has side $\sqrt{2}/2$, no square of side less than $\sqrt{2}$ is capable of accommodating the acceptable set

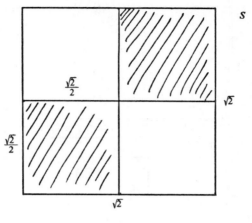

FIG. 75.

consisting of two squares of area $1/2$ (FIG. 75). Consequently, a "universal" square container for the prescribed sets must have side at least $\sqrt{2}$.

Let the squares of a given set be arranged in order of nonincreasing magnitude. Let the lengths of their sides be denoted

$$s_1 \geqslant s_2 \geqslant s_3 \geqslant \cdots,$$

and let us refer to the square of side s_i simply as the square s_i. Let the sides of the proposed container S be placed horizontally and vertically. Starting in the bottom left corner, pack in the squares $s_1, s_2, \ldots,$ in that order, until no more will fit along the bottom. Let the upper edge of the first and largest square (s_1) be extended across S to provide the top of a layer for the row just packed (see FIG. 76). Now start a second row on top of the first one, beginning again at the left, continuing to take the squares as they come in their nonincreasing order. Pack as many as will fit along this second row and again extend the upper edge of the first one in the row to close a layer over them. We shall see that, continuing row on row like this, the entire set can be packed into the square S.

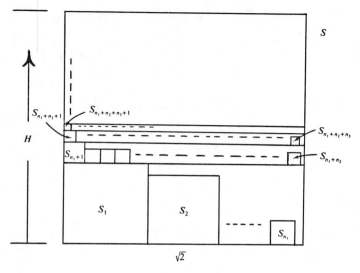

FIG. 76.

Suppose the number of squares in the rth row is n_r. Then the breakdown of the rows is as follows:

row 1 has squares $s_1, s_2, \ldots, s_{n_1}$;

row 2 has squares $s_{n_1+1}, s_{n_1+2}, \ldots, s_{n_1+n_2}$;

row 3 has squares $s_{n_1+n_2+1}, s_{n_1+n_2+2}, \ldots, s_{n_1+n_2+n_3}$;

. . . .

We show that the packing, carried out in this manner, never goes out the top of S, implying that S is able to accommodate all the squares. We do this by proving that the total height H of the stack of rows, measured up the left edge of S, never exceeds $\sqrt{2}$:

$$H = s_1 + s_{n_1+1} + s_{n_1+n_2+1} + \cdots \leqslant \sqrt{2}.$$

Since the total area of the squares is 1, we have

$$1 \geqslant s_1 \geqslant s_2 \geqslant \cdots,$$

showing that S can certainly accommodate the first row. If one row suffices, then there is no problem. Consider therefore the general case in which more than one row is required. Of course, we need to use the fact that the total area is 1. We begin, then, by deriving some relations involving the s_i^2.

Since the first square of the second row is placed there because it is too large to go at the end of the first row, we have

$$s_1 + s_2 + \cdots + s_{n_1} + s_{n_1+1} > \sqrt{2} ,$$

and

$$s_2 + \cdots + s_{n_1+1} > \sqrt{2} - s_1.$$

Multiplying this by s_{n_1+1}, we get

$$s_2 \cdot s_{n_1+1} + s_3 \cdot s_{n_1+1} + \cdots + s_{n_1+1} \cdot s_{n_1+1} > \left(\sqrt{2} - s_1\right) \cdot s_{n_1+1}.$$

Since $s_2 \geqslant s_3 \geqslant s_4 \geqslant \cdots \geqslant s_{n_1+1}$, we have

$$s_2^2 + s_3^2 + \cdots + s_{n_1+1}^2$$
$$= s_2 \cdot s_2 + s_3 \cdot s_3 + \cdots + s_{n_1+1} \cdot s_{n_1+1}$$
$$\geqslant s_2 \cdot s_{n_1+1} + s_3 \cdot s_{n_1+1} + \cdots + s_{n_1+1} \cdot s_{n_1+1}$$
$$> \left(\sqrt{2} - s_1\right) \cdot s_{n_1+1},$$

that is,

$$\boxed{s_2^2 + s_3^2 + \cdots + s_{n_1+1}^2 > \left(\sqrt{2} - s_1\right) \cdot s_{n_1+1}}.$$

Similarly for the second row, we have

$$s_{n_1+1} + s_{n_1+2} + \cdots + s_{n_1+n_2} + s_{n_1+n_2+1} > \sqrt{2} ,$$

and

$$s_{n_1+2} + \cdots + s_{n_1+n_2} + s_{n_1+n_2+1} > \sqrt{2} - s_{n_1+1}.$$

Multiplying by $s_{n_1+n_2+1}$, and noting that $s_{n_1+2} \geqslant \cdots \geqslant s_{n_1+n_2+1}$, we obtain in precisely the same way as for the first row,

$$s_{n_1+2}^2 + s_{n_1+3}^2 + \cdots + s_{n_1+n_2+1}^2 > \left(\sqrt{2} - s_{n_1+1}\right) \cdot s_{n_1+n_2+1}.$$

Because $s_1 \geqslant s_{n_1+1}$, from this we obtain that

$$\boxed{s_{n_1+2}^2 + s_{n_1+3}^2 + \cdots + s_{n_1+n_2+1}^2 > \left(\sqrt{2} - s_1\right) \cdot s_{n_1+n_2+1}}.$$

In a similar way, for the third row we get that

$$\boxed{s_{n_1+n_2+2}^2 + \cdots + s_{n_1+n_2+n_3+1}^2 > (\sqrt{2} - s_1) \cdot s_{n_1+n_2+n_3+1}},$$

and so on for all rows.... Adding these inequalities, we have altogether that

$$s_2^2 + s_3^2 + \cdots > (\sqrt{2} - s_1)(s_{n_1+1} + s_{n_1+n_2+1} + \cdots).$$

Since the total area is 1 and the second factor on the right is merely $H - s_1$, we obtain

$$1 - s_1^2 > (\sqrt{2} - s_1)(H - s_1),$$

$$H - s_1 < \frac{1 - s_1^2}{\sqrt{2} - s_1},$$

and

$$H < \frac{1 - s_1^2}{\sqrt{2} - s_1} + s_1.$$

Now, an easy simplification shows that

$$\sqrt{2} - \frac{(1 - \sqrt{2} \cdot s_1)^2}{\sqrt{2} - s_1} = \frac{1 - s_1^2}{\sqrt{2} - s_1} + s_1.$$

Hence we see that

$$H < \sqrt{2} - \frac{(1 - \sqrt{2} \cdot s_1)^2}{\sqrt{2} - s_1} \leqslant \sqrt{2},$$

giving $H < \sqrt{2}$, as desired.

RED AND GREEN BALLS*

There are 6 red balls and 8 green balls in a bag. Five balls are drawn out at random and placed in a red box; the remaining 9 balls are put in a green box. What is the probability that the number of red balls in the green box plus the number of green balls in the red box is *not* a prime number?

Solution:

Let g denote the number of green balls in the red box. Since there are a total of 6 red balls and 8 green ones, the colors must be distributed over the boxes as illustrated (FIG. 77). Therefore the number of red balls in the green box plus the number of green balls in the red box $=(g+1)+g=$ the odd number $2g+1$. Now g cannot exceed 5, the total number of balls in the red box. Thus $1 \leqslant 2g+1 \leqslant 11$.

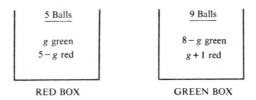

5 Balls		9 Balls
g green		$8-g$ green
$5-g$ red		$g+1$ red
RED BOX		GREEN BOX

FIG. 77.

*AMM, 1960, p. 698, Problem E1400, proposed by S. D. Pratico, Iona College, New Rochelle, New York, solved by C. W. Trigg, Los Angeles City College.

The only odd composite number in this range is 9. However, we must also include the number 1, which is neither prime nor composite. In order to obtain a nonprime value, then, $2g+1$ must be either 1 or 9, making g either 0 or 4. The probability of making a draw with $g=0$ or 4 is

$$\frac{(\#\text{ of ways of taking all 5 red})+(\#\text{ of ways of 4 green and 1 red})}{\text{total number of draws}}$$

$$= \frac{\binom{6}{5}+\binom{8}{4}\binom{6}{1}}{\binom{14}{5}} = \frac{6+420}{2002} = \frac{213}{1001}.$$

COMPOSITE TERMS IN ARITHMETIC PROGRESSION*

In problem #54, page 136, we saw that, in the sequence of natural numbers, there occur arbitrarily long intervals of consecutive numbers all of which are composite. In this section we prove that there exist arbitrarily long arithmetic progressions whose terms are relatively prime in pairs and which consist entirely of composite numbers.

Solution:

We show that, for all natural numbers $n > 1$, there exist n composite numbers in arithmetic progression which are relatively prime in pairs.

Choose any prime number p which exceeds the specified number n. Next, form the possibly large number $p + (n-1)n!$ and select an integer N such that

$$N > p + (n-1)n!.$$

We claim, then, that the n (very likely large) numbers

$$N! + p, N! + p + n!, N! + p + 2 \cdot n!, \ldots, N! + p + (n-1)n!$$

satisfy the required conditions. Clearly they are in arithmetic progression with common difference $n!$. Now, we chose N such that $N > p + (n-1)n!$. Therefore, for $i = 0, 1, 2, \ldots, n-1$, we have

---

*AMM, 1969, p. 199, Problem E2062, proposed by Dale Peterson, Student, Mira Loma High School, Sacramento, California, solved by Arne Garness, Charles Heuer, and Gerald Heuer, Concordia College.

$p + i \cdot n! < N$, implying that the number $(p + i \cdot n!)$ is one of the factors of $N!$. Thus $N! + p + i \cdot n!$, $i = 0, 1, \ldots, n - 1$, is divisible by the lesser $(p + i \cdot n!)$ (which exceeds 1), and is therefore composite.

Also, suppose that some pair of terms

$$N! + p + i \cdot n! \quad \text{and} \quad N! + p + j \cdot n!, i > j,$$

possess a common *prime* divisor q. Then q divides their difference $(i - j)n!$. Now $|i - j| < n$. Therefore every prime divisor of $(i - j)n!$ is $\leqslant n$. Accordingly, $q \leqslant n$. Thus $q | n!$. But $N > n$, and q therefore divides the large $N!$. Dividing $N! + p + i \cdot n!$, then, q must also divide p. But this is impossible since p and q are both primes and $q \leqslant n < p$. The conclusion follows.

ABUTTING EQUILATERAL TRIANGLES*

Equilateral triangles of sides $1, 3, 5, \ldots, 2n-1, \ldots$, are placed end-to-end along a straight line (FIG. 78). Show that the vertices which do not lie on the line all lie on a parabola and that their focal radii are all integers.

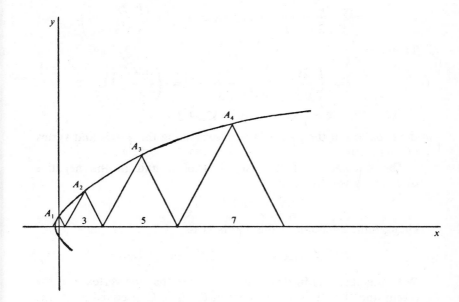

FIG. 78.

*AMM, 1922, p. 92, Problem 2866, proposed by Norman Anning, University of Michigan, solved by G. W. Smith, University of Kansas.

Solution:

Let the vertices in question be denoted, in order, $A_1, A_2, \ldots$, and let the coordinate axes be placed so that the given line is the x-axis and the y-axis passes through A_1. Then A_1 is the point $(0, \sqrt{3}/2)$. Now the base of the nth triangle is $2n-1$. Therefore the x-coordinate of A_n is

$$x = \tfrac{1}{2} + 3 + 5 + \cdots + (2n-3) + \tfrac{1}{2}(2n-1),$$

which reduces easily to $x = n(n-1)$. The y-coordinate of A_n is simply $2n-1$ times that of A_1:

$$y = (2n-1) \cdot \frac{\sqrt{3}}{2}.$$

From this we obtain

$$n = \tfrac{1}{2}\left(\frac{2y}{\sqrt{3}} + 1 \right), \quad \text{which yields } n-1 = \tfrac{1}{2}\left(\frac{2y}{\sqrt{3}} - 1 \right).$$

Therefore

$$x = \tfrac{1}{2}\left(\frac{2y}{\sqrt{3}} + 1 \right) \cdot \tfrac{1}{2}\left(\frac{2y}{\sqrt{3}} - 1 \right) = \tfrac{1}{4}\left(\frac{4y^2}{3} - 1 \right),$$

$$12x = 4y^2 - 3, \quad \text{or} \quad 4y^2 = 12x + 3,$$

the equation of the parabola with axis along the x-axis and vertex at $(-1/4, 0)$.

The translation of the axes $1/4$ of a unit in the negative direction, given by

$$X = x + \tfrac{1}{4}, \, Y = y,$$

changes the equation of the parabola to

$$4Y^2 = 12\left(X - \tfrac{1}{4} \right) + 3 = 12X, \quad \text{or} \quad Y^2 = 3X.$$

With the parabola in this position, we see that the vertex is at the origin and that the focus occurs at $(3/4, 0)$. Consequently, in the original position, the parabola must have had its focus at $(1/2, 0)$

(a vertex of the first triangle). We observe, then, that the focal radius of A_n is given by

$$\sqrt{\left[n(n-1)-\tfrac{1}{2}\right]^2+\left[(2n-1)\cdot\frac{\sqrt{3}}{2}\right]^2}$$
$$=n^2-n+1, \quad \text{an integer.}$$

THE TESTS*

Three students, A, B, C, compete in a series of tests. For coming first in a test, one is awarded x points; for coming second, y points; for coming third, z points. Here x, y, and z are natural numbers and $x > y > z$. There were no ties in any of the tests. Altogether A accumulated 20 points, B 10 points, and C 9 points. Student A came in second in the algebra test. Who came in second in the geometry test?

Solution:

Altogether a total of $20 + 10 + 9 = 39$ points were scored. Since x, y, and z are different natural numbers, at least $1 + 2 + 3 = 6$ points must be awarded for each test. And $x + y + z$ must divide the grand total 39. Because two different tests are noted in the problem, we have $x + y + z \neq 39$. Since the only other divisors of 39 are 1, 3, and 13, and the divisor $x + y + z \geqslant 6$, we must have

$$x + y + z = 13,$$

implying that 3 tests were held.

Since A came second in the algebra test, one part of his score is a y. If A were also to have a z in his score, then the best he could do is have the remaining part an x, giving him a total of $y + z + x = 13$. Since A's total is 20, there can be no z in his score. Consequently, A's score of 20 is either $3y$, $x + 2y$, or $2x + y$. Since $3 \nmid 20$, it cannot be $3y$. Now, if his score were $x + 2y$, then we

---

*Derived from Problem 1 on the 16th International Olympiad, 1974.

would have

$$x + 2y = x + y + y = 20,$$

while

$$x + y + z = 13.$$

Subtraction yields $y - z = 7$. Since $x > y > z$, this means $y \geqslant 8$ and $x \geqslant 9$. Then $x + y \geqslant 17$, contradicting $x + y + z = 13$. Thus A's score must be $2x + y = 20$.

Because of this, y must be an even number. Since $y \geqslant 6$ implies $x \geqslant 7$, and makes $x + y \geqslant 13$, leaving nothing for z in $x + y + z = 13$, the value of y must be either 2 or 4.

For $y = 2$, we must have the lesser $z = 1$, implying $x = 10$ in order to make $x + y + z = 13$. In this case, however, A's score of $2x + y = 22$, not 20. Therefore $y = 4$, and from $2x + y = 20$, we get $x = 8$. From $x + y + z = 13$ we have $z = 1$.

There is only one way these values can go together to yield accumulated scores of 20, 10 and 9:

	(i)	(ii)	(iii)	total
A	8	8	4	20
B	1	1	8	10
C	4	4	1	9

Since C is second whenever A isn't, C must have been second in the geometry test.

AN APPLICATION OF PTOLEMY'S
THEOREM

Ptolemy's theorem states that, for a cyclic quadrilateral, the product of the lengths of the diagonals is equal to the sum of the products of the lengths of the pairs of opposite sides (for a proof, see almost any text in synthetic geometry). This provides an immediate proof of the following result:

Let $A_1 A_2 A_3$ denote an equilateral triangle inscribed in a circle. For any point P on the circle, show that the two shorter segments among PA_1, PA_2, PA_3 add up to the third one.

Proof. Let s denote the length of the side of the given triangle. Referring to FIG. 79, by Ptolemy's theorem we have

$$s \cdot PA_2 = s \cdot PA_1 + s \cdot PA_3,$$

and

$$PA_2 = PA_1 + PA_3.$$

In this section we consider two generalizations of this proposition.

(i) (*AMM*, 1933, *p.* 501, *Problem* 3583, *proposed by H. Grossman, New York, solved by Laurence Hadley, Purdue University.*)

Let $A_0 A_1 \cdots A_{3n-1}$ denote a regular $3n$-gon inscribed in a circle. From any point P on the circle, chords are drawn to the $3n$ vertices (FIG. 80). Prove that the sum of the n longest chords is the same as the sum of the $2n$ shortest ones.

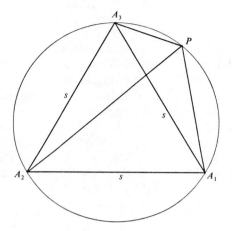

FIG. 79.

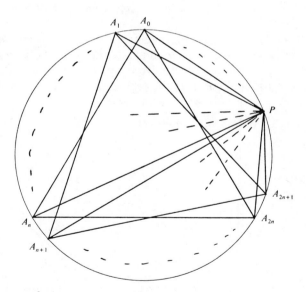

FIG. 80.

Solution:

The vertices of a regular $3n$-gon go together in 3's to determine n equilateral triangles inscribed in the circle. For the three chords associated with each of these triangles, the lengths of the shorter two add up to the length of the third. We observe that the lengths of the shorter two chords of such a set are each $\leqslant s$ (the length of the side of the equilateral triangle), while the length of the longest chord is $\geqslant s$. Taking the longest chord from each three, then, we obtain the n longest chords of the entire set, and the remaining chords constitute the $2n$ shortest ones. By the fundamental result above, the sum of the lengths of the n longest chords is the same as that of the other $2n$.

(ii) (*The following result was established for $n=3$ and 5 by William Wernich, City College, New York; the general case was stated and established by Leroy Dickey, University of Waterloo.*)

Suppose $A_1, A_2, \ldots, A_n$ is a regular n-gon, where n is o*dd*, and P is an arbitrary point on its circumcircle. Let the labels be assigned so that P occurs on the closed arc $A_n A_1$. Let $PA_i = a_i$. Then, if we alternately add and subtract the segments radiating from P,

$$a_1 - a_2 + a_3 - a_4 + \cdots - \cdots + a_n,$$

we invariably obtain the value 0 (FIG. 81).

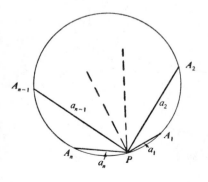
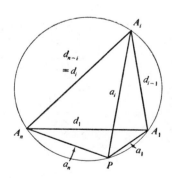

FIG. 81.

Solution:

For $n=3$, this is precisely our fundamental result, above. Whereas one application of Ptolemy's theorem was sufficient in that case, we use it repeatedly in proving this generalization.

Let d_i denote the length of a diagonal of the n-gon which cuts off i edges on one side and $n-i$ edges on the other. Then $d_i = d_{n-i}$ for all i. Note that d_1 is merely a side of the n-gon. Thus the sides and diagonals of quadrilateral $A_n PA_1 A_i$ are as shown in the diagram. By Ptolemy's theorem, we have

$$d_1 a_i = d_i a_1 + d_{i-1} a_n.$$

Putting $i = 2, 3, \ldots, n-1$, we obtain n-2 results which, preceded by the identity $d_1 a_1 = d_1 a_1$ and followed by $d_1 a_n = d_{n-1} a_n$, yield the set of n equations

$$d_1 a_1 = d_1 a_1$$
$$d_1 a_2 = d_2 a_1 + d_1 a_n$$
$$d_1 a_3 = d_3 a_1 + d_2 a_n$$
$$d_1 a_4 = d_4 a_1 + d_3 a_n$$
$$\cdots \cdots \cdots \cdots \cdots \cdots$$
$$d_1 a_{n-1} = d_{n-1} a_1 + d_{n-2} a_n$$
$$d_1 a_n = d_{n-1} a_n.$$

Alternately adding and subtracting from the top, we obtain

$$d_1(a_1 - a_2 + \cdots - \cdots + a_n)$$
$$= (a_1 - a_n)(d_1 - d_2 + d_3 - d_4 + \cdots - \cdots + d_{n-2} - d_{n-1})$$

(it is $-d_{n-1}$ because $n-1$ is even). But $d_1 = d_{n-1}$, $d_2 = d_{n-2}$, and so on, making $d_1 - d_2 + d_3 - d_4 + \cdots - \cdots + d_{n-2} - d_{n-1} = 0$. Therefore

$$a_1 - a_2 + a_3 - a_4 + \cdots - \cdots + a_n = 0.$$

ANOTHER DIOPHANTINE EQUATION*

If y and z are natural numbers satisfying

$$y^3 + 4y = z^2,$$

prove that y is twice a square.

Solution:

Let k^2 denote the greatest square which divides y and suppose that $y = nk^2$. Then n cannot have any repeated factors, lest a square greater than k^2 divide y. Then

$$y^3 + 4y = z^2$$

gives

$$y(y^2 + 4) = z^2,$$
$$nk^2(y^2 + 4) = z^2,$$

implying that

$$k^2 | z^2 \quad \text{and thus} \quad k | z.$$

Let $z = mk$. Then $nk^2(y^2 + 4) = m^2 k^2$ and $n(y^2 + 4) = m^2$. That is to say, $n(y^2 + 4)$ is a perfect square. But n has no repeated factors. Thus all the factors of n must occur again in $y^2 + 4$. This means that

$$n | y^2 + 4.$$

---

*AMM, 1973, p. 77, Problem E2332, proposed by R. S. Luthar, University of Wisconsin at Janesville, solved by G. B. Robinson, SUNY at New Paltz.

Since $y = nk^2$, we have

$$n \mid n^2 k^4 + 4, \quad \text{and} \quad n \mid 4.$$

Thus $n = 1$, 2, or 4. Since n has no repeated factors, $n \neq 4$. If n were to be 1, then, from $n(y^2 + 4) = m^2$, we get

$$y^2 + 4 = m^2.$$

But no two squares differ by 4. Hence n must be 2, if there are any solutions at all, making y twice a square.

We note that $y = 2(= 2 \cdot 1^2)$, $z = 4$ is a solution. Indeed it can be shown that this is the only solution in natural numbers.

AN UNUSUAL PROPERTY
OF COMPLEX NUMBERS*

It is true that, for $x = 1 + i\sqrt{3}$, $y = 1 - i\sqrt{3}$, and $z = 2$, where $i = \sqrt{-1}$,

$$x^5 + y^5 = z^5, \quad x^7 + y^7 = z^7, \quad \text{and} \quad x^{11} + y^{11} = z^{11}.$$

Prove the surprising generalization that, for this particular choice of $x, y,$ and z, the equation

$$x^p + y^p = z^p$$

holds for every prime $p > 3$.

Solution:

An easy calculation shows that both x^6 and y^6 are equal to 2^6. Since $6n, 6n+2, 6n+3,$ and $6n+4$ are never primes, a prime number > 3 must have the form $6n+1$ or $6n-1$.

For $p = 6n + 1$, we have

$$
\begin{aligned}
x^p + y^p &= x^{6n+1} + y^{6n+1} \\
&= (x^6)^n \cdot x + (y^6)^n \cdot y \\
&= 2^{6n} \cdot x + 2^{6n} \cdot y \\
&= 2^{6n}(x + y).
\end{aligned}
$$

*AMM, 1943, p. 63, Problem E518, proposed and solved by J. Rosenbaum, Bloomfield, Connecticut.

But $x + y = 2$, and we get

$$x^p + y^p = 2^{6n} \cdot 2 = 2^{6n+1} = z^p.$$

Observing that

$$\frac{1}{x} + \frac{1}{y} = \frac{1}{2} = \frac{1}{z},$$

we can similarly show that the equation holds for $p = 6n - 1$.

A CHAIN OF CIRCLES*

A circle C_0 of radius 1 km is tangent to a line L at Z (FIG. 82). A circle C_1 of radius 1 mm is drawn tangent to both C_0 and L, on the right-hand side of C_0. A family of circles C_i is constructed outwardly to the right side so that each C_i is tangent to C_0, L, and to the previous circle C_{i-1}. Eventually the members become so large that it is impossible to enlarge the family any further. How many circles can be drawn before this happens?

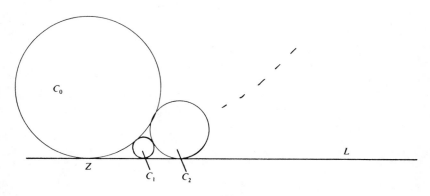

FIG. 82.

*This problem comes from the book Mathematical Games and Pastimes, by A. P. Domoryad, Pergamon Press, 1964, problem 19, p. 242; the solution was received from C. Stanley Ogilvy in private communication, 1974.

Solution:

Let distances be expressed in mm, making the radius of $C_1 = 1$ and that of $C_n = 10^6$. Let the chain of circles be inverted in the circle I which has center Z and radius $2 \cdot 10^6$ (FIG. 83). (For an account of circular inversion, see the reference.) Since L goes through the center of inversion Z, it inverts into itself. Now I and C_0 are tangent internally, say at X. Since C_0 goes through Z, its image C_0' is a straight line, namely the common tangent to I and C_0 (at X). Then L and C_0' are the tangents to C_0 at opposite ends of a diameter. Therefore they are parallel and determine a strip S in the plane.

Since each C_i in the chain is tangent to C_0 and L, each image C_i' is tangent to the edges of S. Since none of these circles, except C_0, goes through Z, the images C_i' are circles. Consequently, the images of the pairwise tangent circles in the chain constitute a row of equal touching circles bounded by the edges of S. All the image circles are the same size as C_0. Since C_1 is inside the circle of inversion I near the center Z, its image is outside I far along the strip from I. As i runs through $1, 2, 3, \ldots$, the row of image circles works its way back along S toward C_0.

Suppose C_1 and C_1' touch L at Y and Y', respectively. It is an easy application of the theorem of Pythagoras to determine that

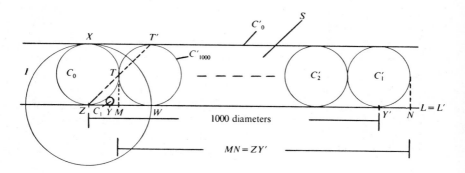

FIG. 83.

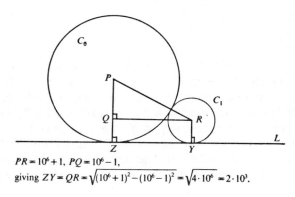

$PR = 10^6 + 1$, $PQ = 10^6 - 1$,

giving $ZY = QR = \sqrt{(10^6 + 1)^2 - (10^6 - 1)^2} = \sqrt{4 \cdot 10^6} = 2 \cdot 10^3$.

FIG. 84.

$ZY = 2 \cdot 10^3$ (FIG. 84). The inversion relation, then, yields

$$ZY \cdot ZY' = (2 \cdot 10^6)^2$$
$$2 \cdot 10^3 (ZY') = 4 \cdot 10^{12}$$
$$ZY' = 2 \cdot 10^9 = 1000(2 \cdot 10^6),$$

which is 1000 diameters of the C_i'.

Consequently, there is room for exactly 1000 circles in the row of image circles from C_1' to C_0. That is to say, C_{1000}' just touches C_0.

Now it is not difficult to see that C_{1000}' is its own image under inversion in I. Since I has twice the radius of each of C_0 and C_{1000}', which are tangent at T, the image T' on C_0', being in line with Z and T, is directly above W, the point of contact of C_{1000}' and L, which is also the point where I crosses L. Thus the inversion takes the three points T, T', and W of C_{1000}' respectively into T', T, and W, implying that C_{1000}' is carried into itself. As a result, the circle $C_{1000} \equiv C_{1000}'$ is the same size as C_0. In constructing the chain of circles, then, the circles have grown to be the same size as the original C_0 by the time one reaches C_{1000}. Therefore it is impossible to extend the chain beyond this point, and C_{1000} is the last circle.

(We observe that our family of circles constitutes part of a Steiner chain, relative to C_0 and L, considering L to be a circle of infinite radius.)

Reference

Coxeter and Greitzer, Geometry Revisited, vol. 19, New Mathematical Library, Math. Assoc. of America, p. 108 ff.

REPEATED DIGITS AT THE END
OF A SQUARE*

Determine the number of digits in the longest string of repeated nonzero digits in which a perfect square can end and find the least square which ends in such a maximum sequence.

Solution:

For every natural number n, we have

$$n \equiv 0, \pm 1, \pm 2, \pm 3, \pm 4, \quad \text{or} \quad 5 \pmod{10},$$

giving

$$n^2 \equiv 0, 1, 4, 9, 6, \quad \text{or} \quad 5 \pmod{10}.$$

Therefore no square ends in $2, 3, 7,$ or 8. We need deal, then, only with the digits $1, 4, 5, 6,$ and 9. Now n is either even or odd. Hence

$$n^2 = (2a)^2 = 4a^2, \quad \text{or} \quad n^2 = (2a+1)^2 = 4(a^2+a)+1,$$

showing $n^2 \equiv 0$ or $1 \pmod 4$. Since the number

$$ab \cdots cxy = ab \cdots c00 + xy \equiv xy \pmod 4,$$

we see that it is impossible for a square to end in $11, 55, 66,$ or 99 because each of these is $\equiv 2$ or $3 \pmod 4$. Thus the only possibility for a repeated nonzero digit at the end of a square is 4.

If a square ends in as many as four 4's, we have

$$n^2 = ab \ldots c4444 = ab \ldots c0000 + 4400 + 44.$$

*A problem from the 1970 Putnam Examination.

Since $16|10000$ and $16|4400$, this gives $n^2 \equiv 12$ (mod 16). But, modulo 16, we have $n \equiv 0, \pm 1, \pm 2, \pm 3, \pm 4, \pm 5, \pm 6, \pm 7,$ or 8, and

$$n^2 \equiv 0, 1, 4, 9, 0, 9, 4, 1, \quad \text{or} \quad 0, \quad \text{but never 12.}$$

Hence a square can end in no more than 3 repeated nonzero digits. Since 444 is not a square, the discovery that

$$1444 = 38^2$$

shows that it is possible for a square to end in 3 4's and exhibits the least such square.

AN ANGLE BISECTOR*

In $\triangle ABC$, $AC = BC$. A circle K is drawn with center C and radius $< AC$ (Fig. 85). Find a point P on K at which the tangent bisects angle APB.

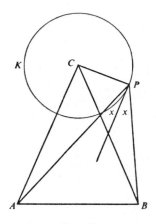

FIG. 85.

Solution:

The surprise here is that, for all circles K, the point P lies on the circumcircle S of $\triangle ABC$ (Fig. 86).

*AMM, 1927, p. 102, Problem 3170, proposed by A. A. Bennett, Lehigh University, solved by Velma Maness, Oklahoma University.

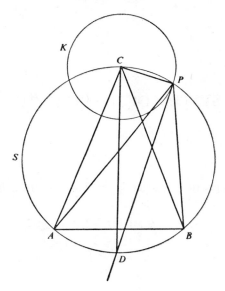

FIG. 86.

Let *CD* be the diameter of *S* through *C*. Then $\angle CPD$ is a right angle, making *PD* the tangent to *K* at *P*. But, since $AC = BC$, *CD* bisects $\angle C$ and therefore *D* bisects arc *AB*. Thus $\angle APD = \angle DPB$.

A SYSTEM OF INEQUALITIES*

What is the greatest integer n for which there exists a simultaneous solution x to the inequalities

$$k < x^k < k+1, \quad k = 1, 2, 3, \dots, n:$$
$$1 < x < 2$$
$$2 < x^2 < 3$$
$$3 < x^3 < 4$$
$$4 < x^4 < 5$$
$$\cdots \cdots ?$$

Solution:

The greatest possible n is 4. If some x were to satisfy as many as the first five of these inequalities, then, from the third one we would have

$$3 < x^3,$$

and from the fifth one, $x^5 < 6$. These yield

$$3^5 < x^{15} < 6^3,$$

which claims that $243 < 216$. Thus $n \leqslant 4$.

Since $\sqrt[4]{4} = \sqrt{2}$, any x between $\sqrt[3]{3}$ and $\sqrt[4]{5}$ satisfies the first four inequalities.

*AMM, 1960, p. 476, Problem E1388, proposed by H. W. Gould, West Virginia University, solved by N. J. Fine, Institute for Advanced Study.

AN UNEXPECTED PROPERTY OF A
REGULAR 26-GON*

In a circle, center O, a regular 26-gon $A_1A_2....A_{26}$ is inscribed (FIG. 87). The center O is reflected in chords $A_{25}A_1$ and A_2A_6 to give images O_1 and O_2, respectively. Prove the remarkable property that O_1O_2 is the side of an equilateral triangle inscribed in the circle.

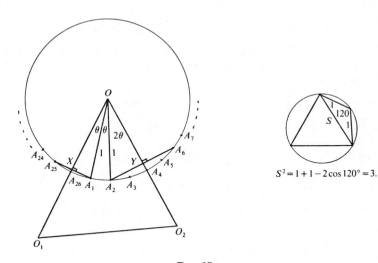

$$S^2 = 1 + 1 - 2\cos 120° = 3.$$

FIG. 87.

*AMM, 1958, p. 716, Problem 4768, proposed by Victor Thébault, Tennie, Sarthe, France, solved by W. J. Blundon, University of Newfoundland.

Solution:

Let the radius of the circle be taken as the unit of length. We observe that an equilateral triangle is inscribed in a circle by joining alternate vertices of an inscribed regular hexagon, which is obtained by stepping the radius around the circumference. From the law of cosines, we find that the side of an equilateral triangle in a unit circle is $\sqrt{3}$. We show, then, that $O_1O_2 = \sqrt{3}$.

Because OO_1 is the perpendicular bisector of $A_{25}A_1$, it passes through A_{26}. Similarly, OO_2 goes through A_4. Let the angle subtended at the center by a side of the 26-gon be denoted θ. Then $\theta = 2\pi/26 = \pi/13$. Thus $\angle A_1OA_2 = \theta$; and if the midpoints of $A_{25}A_1$ and A_2A_6 are X and Y, respectively, we have $\angle A_1OX = \theta$ and $\angle A_2OY = 2\theta$. Since radii A_1O, A_2O are 1, we have $OX = \cos\theta$, $OY = \cos 2\theta$, and their doubles $OO_1 = 2\cos\theta$, $OO_2 = 2\cos 2\theta$.

The law of cosines applied to $\triangle OO_1O_2$ gives

$$O_1O_2^2 = OO_1^2 + OO_2^2 - 2(OO_1)(OO_2)\cos 4\theta$$

$$= 4\cos^2\theta + 4\cos^2 2\theta - 2(2\cos\theta)(2\cos 2\theta)(\cos 4\theta).$$

The argument finishes with the use of only well-known trigonometric formulas. From $\cos 2x = 2\cos^2 x - 1$, we get $4\cos^2 x = 2 + 2\cos 2x$, and then

$$O_1O_2^2 = (2 + 2\cos 2\theta) + (2 + 2\cos 4\theta) - 8\cos\theta\cdot\cos 2\theta\cdot\cos 4\theta$$

$$= 4 + 2\cos 2\theta + 2\cos 4\theta - 8\cos\theta\cdot\cos 2\theta\cdot\cos 4\theta.$$

Surprisingly, we now multiply through by $\sin\theta$:

$$O_1O_2^2\sin\theta = 4\sin\theta + 2\sin\theta\cdot\cos 2\theta + 2\sin\theta\cdot\cos 4\theta$$
$$- 8\sin\theta\cdot\cos\theta\cdot\cos 2\theta\cdot\cos 4\theta.$$

But

$$8\sin\theta\cdot\cos\theta\cdot\cos 2\theta\cdot\cos 4\theta = 4\sin 2\theta\cdot\cos 2\theta\cdot\cos 4\theta$$
$$= 2\sin 4\theta\cdot\cos 4\theta = \sin 8\theta.$$

And, for all natural numbers n.

$$2\sin\theta\cdot\cos n\theta = \sin(n+1)\theta - \sin(n-1)\theta.$$

Therefore

$$O_1 O_2^2 \sin\theta = 4\sin\theta + (\sin 3\theta - \sin\theta)$$
$$+ (\sin 5\theta - \sin 3\theta) - \sin 8\theta.$$

However, $\theta = \pi/13$, implying $13\theta = \pi$, or $5\theta + 8\theta = \pi$, making $\sin 5\theta = \sin 8\theta$. Therefore

$$O_1 O_2^2 \sin\theta = 3\sin\theta, \quad \text{and} \quad O_1 O_2 = \sqrt{3}.$$

MORE ON PERFECT SQUARES*

Prove that the only integers x which make

$$x^4 + x^3 + x^2 + x + 1$$

a perfect square are $x = -1, 0,$ and 3.

Solution:

(This problem went unsolved for 7 years until Professor Bennett came up with the following clever attack based on the familiar notion of "completing the square".)

We seek integral solutions of the equation

$$y^2 = x^4 + x^3 + x^2 + x + 1.$$

Let us try to determine y in terms of x. We see that

$$\left(x^2 + \frac{x}{2}\right)^2 = x^4 + x^3 + \frac{x^2}{4},$$

providing the correct first two terms. The value of $x^2 + \frac{x}{2} + 1$ is pretty close; we have

$$\left(x^2 + \frac{x}{2} + 1\right)^2 = x^4 + x^3 + \frac{9}{4}x^2 + x + 1 = y^2 + \frac{5}{4}x^2,$$

which is too great unless $x = 0$. On the other hand,

$$\left(x^2 + \frac{x}{2} + \frac{\sqrt{5} - 1}{4}\right)^2 = x^4 + x^3 + \frac{2\sqrt{5} - 1}{4}x^2 + \frac{\sqrt{5} - 1}{4}x + \frac{3 - \sqrt{5}}{8}$$

$$= y^2 - \frac{5 - 2\sqrt{5}}{4}\left(x + \frac{3 + \sqrt{5}}{2}\right)^2 < y^2.$$

*AMM, 1926, p. 281, Problem 2784, proposed by T. H. Gronwall, New York City, solved by A. A. Bennett, Lehigh University.

because $(5-2\sqrt{5})/4$ is positive and $x \neq -(3+\sqrt{5})/2$, an irrational number. Thus we have estimates which are too large and too small. The magnitude of y, then, must lie somewhere in the range determined by these approximations:

$$x^2 + \frac{x}{2} + \frac{\sqrt{5}-1}{4} < |y| \leqslant x^2 + \frac{x}{2} + 1,$$

that is,

$$x^2 + \frac{x + \left(\dfrac{\sqrt{5}-1}{2}\right)}{2} < |y| \leqslant x^2 + \frac{x+2}{2}.$$

For some real number k in the range $(\sqrt{5}-1)/2 < k \leqslant 2$, the magnitude of y must be given exactly by

$$|y| = x^2 + \frac{x+k}{2}.$$

But x and y are whole numbers. Thus $(x+k)/2$ must be an integer, making $x+k$ an even integer. Then k, also, must be an integer. But the only integers in the range for k are 1 and 2 (since the value of $(\sqrt{5}-1)/2$ lies between 0 and 1). Thus $k = 1$ or 2.

For $k = 2$, we have

$$|y| = x^2 + \frac{x}{2} + 1,$$

and from

$$\left(x^2 + \frac{x}{2} + 1\right)^2 = y^2 + \frac{5}{4}x^2,$$

obtained earlier, we get $y^2 = y^2 + \frac{5}{4}x^2$, giving $x = 0$.

For $k = 1$, we have

$$|y| = x^2 + \frac{x+1}{2}.$$

However, it is easily verified that

$$y^2 = \left(x^2 + \frac{x+1}{2}\right)^2 - \frac{(x-3)(x+1)}{4}.$$

For $k = 1$, then, we have

$$y^2 = y^2 - \frac{(x-3)(x+1)}{4},$$

implying that $(x-3)(x+1)=0$, giving $x=3$, or -1. The conclusion follows the routine verification that $x=-1$, 0, and 3 actually yield integral values of y.

Having felt for a long time that this nice argument constituted a rather noteworthy achievement, I was very surprised and delighted to discover a similar, but much more elegant, solution in Mathematical Digest, a humble, mimeographed publication for the high school students in the neighborhood of Cape Town, South Africa. The editor is Professor John H. Webb, University of Cape Town. The July 1973 issue carried the following solution.

Observe that

$$\left(x^2+\frac{x}{2}\right)^2 = x^4+x^3+\frac{x^2}{4}$$

$$= x^4+x^3+x^2+x+1-\left(\frac{3x^2}{4}+x+1\right)$$

$$= y^2-\frac{1}{4}(3x^2+4x+4).$$

Because the discriminant of $3x^2+4x+4$ (namely $4^2-4\cdot3\cdot4=-32$) is negative, the value of $3x^2+4x+4$ is positive for all real numbers x. Thus

$$\left(x^2+\frac{x}{2}\right)^2 < y^2.$$

This gives

$$\left|x^2+\frac{x}{2}\right| < |y|.$$

Since $x^2+x/2=x(x+\frac{1}{2})$ is non-negative for all integers x, we have

$$\left|x^2+\frac{x}{2}\right| = x^2+\frac{x}{2},$$

giving

$$x^2+\frac{x}{2} < |y|.$$

Now, if x is an even integer, $x^2+(x/2)$ is a whole number, and the larger integer $|y|$ must exceed it by at least 1. If x is odd, then

$x^2 + (x/2)$ lies halfway between two consecutive whole numbers, and $|y|$ may possibly exceed it by as little as $\frac{1}{2}$. In any case, we have

$$|y| \geqslant \left(x^2 + \frac{x}{2}\right) + \frac{1}{2},$$

giving

$$y^2 \geqslant x^4 + x^3 + \frac{5x^2}{4} + \frac{x}{2} + \frac{1}{4}$$
$$= x^4 + x^3 + x^2 + x + 1 + \left(\frac{x^2}{4} - \frac{x}{2} - \frac{3}{4}\right)$$
$$= y^2 + \frac{1}{4}(x^2 - 2x - 3).$$

This implies

$$x^2 - 2x - 3 \leqslant 0,$$
$$(x+1)(x-3) \leqslant 0,$$

which restricts x to the values $(-1, 0, 1, 2, 3)$. The solution concludes with the elimination of $x = 1$ and 2 by the direct trial of these values.

AN UNUSUAL POLYNOMIAL*

If $P(x)$ is a polynomial of degree n such that
$$P(x) = 2^x \quad \text{for} \quad x = 1, 2, 3, \ldots, n+1, \quad \text{find } P(n+2).$$

Solution:

The binomial theorem gives

$$2^m = (1+1)^m = \binom{m}{0} + \binom{m}{1} + \cdots + \binom{m}{m} \quad \text{for} \quad m = 1, 2, \ldots, n+1, \ldots.$$

Now, n is an arbitrary but definite number. Consider the polynomial

$$f(x) = 2\left[\binom{x-1}{0} + \binom{x-1}{1} + \cdots + \binom{x-1}{n} \right].$$

The term of greatest degree in $f(x)$ is

$$2\binom{x-1}{n} = 2\frac{(x-1)(x-2)\cdots(x-n)}{n!},$$

of degree n in x. Thus $f(x)$ is of degree n. Now, for $x = 1,$ $2, \ldots, n+1$, we have

$$f(x) = 2\left[(1+1)^{x-1} \right] = 2^x \quad \left(\text{for } m < n, \binom{m}{n} = 0 \right).$$

That is, $f(x)$ and the required $P(x)$ are both of degree n and they agree for the $n+1$ values $x = 1, 2, \ldots, n+1$. Thus $f(x)$ is identically

---

*Pi Mu Epsilon, Vol. 4, 1964, p. 77, Problem 158, proposed and solved by Murray Klamkin, University of Minnesota.

equal to $P(x)$, and we have

$$P(n+2) = 2\left[\binom{n+1}{0} + \binom{n+1}{1} + \cdots + \binom{n+1}{n}\right]$$
$$= 2\left[2^{n+1} - \binom{n+1}{n+1}\right] = 2^{n+2} - 2.$$

Similarly, we obtain

$$P(n+3) = 2\left[2^{n+2} - \binom{n+2}{n+1} - \binom{n+2}{n+2}\right]$$
$$= 2\left[2^{n+2} - (n+2) - 1\right] = 2^{n+3} - 2n - 6.$$

CYCLIC CENTROIDS*

A, B, C, D are 4 points on a circle. Let G_A, G_B, G_C, G_D, respectively, denote the centroids of the triangles BCD, ACD, ABD, ABC (FIG. 88). Prove that G_A, G_B, G_C, G_D also lie on a circle.

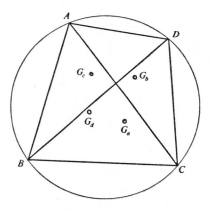

FIG. 88.

Solution:

Let a unit mass be suspended at each of A, B, C, D, and let G denote the center of gravity of this system. Unit masses at A, B, and C are equivalent to a mass of 3 units at G_D. Thus the system is

*AMM, 1965, p. 1026, Problem E1740, proposed by D. P. Ambrose, University of Colorado, solved by Michael Goldberg, Washington D. C.

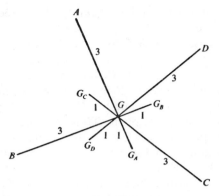

FIG. 89.

equivalent to masses of 3 at G_D and 1 at D. Hence G must lie on $G_D D$ and divide it in the ratio $1:3$ (FIG. 89).

Similarly, G must lie on each of $G_A A$, $G_B B$, and $G_C C$ and divide each in the ratio $1:3$. Consequently, the dilatation $G(-1/3)$ would carry A, B, C, D, respectively, into G_A, G_B, G_C, G_D. But dilatations take circles into circles. Since A, B, C, D lie on a circle, so do G_A, G_B, G_C, G_D.

We observe that, for the n vertices of any polygon, a similar polygon is determined by the n centroids of the sets of $(n-1)$ vertices. Also, a similar condition holds in 3-space.

The "mechanics" approach often provides a smooth solution to geometry problems of considerable difficulty. It is a simple matter to establish the existence of such points in a triangle as the centroid, the incenter, the Gergonne and Nagel points, and to prove such theorems as Ceva's and Menelaus'.

Consider the following problem, which is taken from an elementary contest held in Rumania:

A, B, C are fixed points in a plane and A', B', C' are variable points in another plane π. The midpoints of AA', BB', CC' are L, M, N. What is the locus of the centroid S of $\triangle LMN$?

Let a unit mass be suspended at each of A, B, C, A', B', C'. The masses at A and A' are equivalent to a mass of 2 at the midpoint L. Similarly, we see that the entire system is equivalent to masses of 2 at each of L, M, and N. Consequently, the centroid S of $\triangle LMN$ is the center of gravity of the entire system.

Now, the masses at A, B, C are equivalent to a mass of 3 at G, the centroid of $\triangle ABC$. Similarly, the masses at A', B', C' are equivalent to a mass of 3 at G', the centroid of $\triangle A'B'C'$. As a result, the center of gravity S must bisect the segment GG'. As A', B', C' vary in their plane π, G' takes all positions in π. However, G is fixed in its plane. Thus the locus of S is the plane into which the plane π is carried by the dilatation $G(1/2)$.

Our final problem in this section is due to Murray Klamkin, University of Waterloo:

Prove that the points of contact P, Q, R, S of a skew quadrilateral $ABCD$ which circumscribes a given sphere lie on a circle.

The tangents to a sphere from a point are all the same length. Let $AP = AS = a$, $BP = BQ = b$, $CQ = CR = c$, $DR = DS = d$ (FIG. 90). At A, B, C, D, respectively, suspend masses of $1/a, 1/b, 1/c, 1/d$. The masses at A and B have moments about P

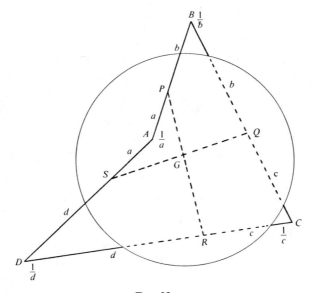

FIG. 90.

which are equal and opposite. Thus they are equivalent to a mass at P of $(1/a)+(1/b)$. Similarly, the masses at C and D are equivalent to some mass at R. Consequently, the center of gravity G of the entire system must occur somewhere on PR. Similarly, pairing the masses at B and C, and also at A and D, we see that G must also lie on QS. The significance of this is that the segments PR and QS must intersect (at G). Consequently, they determine some plane π. Clearly, the points of contact P, Q, R, S belong to the circle of intersection of π and the sphere.

AN EASY REMAINDER*

What is the remainder when $x + x^9 + x^{25} + x^{49} + x^{81}$ is divided by $x^3 - x$?

Solution:

We have

$$\frac{x^{81} + x^{49} + x^{25} + x^9 + x}{x^3 - x} = \frac{x^{80} + x^{48} + x^{24} + x^8 + 1}{x^2 - 1}$$

$$= \frac{(x^{80} - 1) + (x^{48} - 1) + (x^{24} - 1) + (x^8 - 1) + 5}{x^2 - 1}.$$

Since $x^2 - 1$ divides $x^{2n} - 1$, the remainder appears to be 5. Recalling that x was cancelled in the numerator and denominator, the remainder really is $5x$, since

$$\frac{5}{x^2 - 1} = \frac{5x}{x^3 - x}.$$

*AMM, 1973, p. 640, from a Stanford Competitive Mathematics Examination.

A CURIOUS PROPERTY OF 3*

Prove that if m and n are two natural numbers, then one of $\sqrt[n]{m}$, $\sqrt[m]{n}$ is always less than or equal to $\sqrt[3]{3}$.

Solution:

Suppose $m = n$. We wish to show, then, that $\sqrt[n]{n} \leqslant \sqrt[3]{3}$, that is, $n^{1/n} \leqslant 3^{1/3}$, or $n^3 \leqslant 3^n$. This follows easily by induction, as we shall see. Clearly, the proposition is valid for $n = 1, 2$, and 3. Suppose, then, that we have

$$n^3 \leqslant 3^n \quad \text{for some value of } n, \ n \geqslant 3.$$

Then

$$3^{n+1} = 3 \cdot 3^n \geqslant 3n^3 = n^3 + 3n^2 + 3n + (n-3)n^2 + (n^2 - 3)n$$
$$> n^3 + 3n^2 + 3n + 1$$
$$= (n+1)^3.$$

Thus the assertion holds for $m = n$.

This would seem to be only the special case, with $m \neq n$ the general situation. However, things are just the reverse. For, if $m < n$ say, then

$$\sqrt[n]{m} < \sqrt[n]{n} \leqslant \sqrt[3]{3}.$$

We observe that the equality holds in the single case of $m = n = 3$.

*AMM, 1970, p. 768, Problem E2190, proposed by Harry Pollard, Purdue University, solved by Charles Wexler, Arizona State University, and 118 others.

A SQUARE WITHIN A SQUARE*

Lines from the vertices of a square to the midpoints of the sides are drawn, as shown (Fig. 91). Prove the surprising result that the area of the smaller square this produces is $1/5$ the given square.

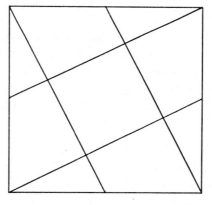

Fig. 91.

Solution:

In a "cross" consisting of 5 equal squares, as shown (Fig. 92) it is evident that the line AD crosses PQ at its midpoint X, and that triangles APX and XQD are congruent. Thus, let $\triangle APX$ be cut off and placed in the position of $\triangle XQD$. By performing this same

*Scripta Mathematica, 1953, p. 270, Curiosa 348, by Nev. R. Mind.

rearrangement at each of A, B, C, and D, it is not difficult to see that we obtain a square.

However, the area is unchanged, implying that the central square still has $1/5$ of the area.

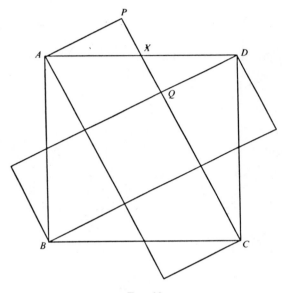

FIG. 92.

ALWAYS A SQUARE*

Write down an even number of 1's to form a number A, and half as many 4's to make a number B. Prove that $A + B + 1$ is always a perfect square.

Solution:

Suppose A has $2m$ digits and B m digits. Then

$$\left(\frac{3B}{4}\right)^2 = \left[\frac{3}{4}\,(\underbrace{44\ldots4}_{m})\right]^2 = (\underbrace{33\ldots3}_{m})\,(\underbrace{33\ldots3}_{m})$$

$$= (\underbrace{11\ldots1}_{m})\,(\underbrace{99\ldots9}_{m}) = (\underbrace{11\ldots1}_{m})\,(10^m - 1)$$

$$= \underbrace{11\ldots1}_{m}\,\underbrace{00\ldots0}_{m} - \underbrace{11\ldots1}_{m}.$$

Adding $\frac{1}{2}B$, which is $\underbrace{22\ldots2}_{m} = 2\,(\underbrace{11\ldots1}_{m})$, we obtain

$$\left(\frac{3B}{4}\right)^2 + \tfrac{1}{2}B = \underbrace{11\ldots1}_{m}\,\underbrace{00\ldots0}_{m} + \underbrace{11\ldots1}_{m} = \underbrace{11\ldots\ldots1}_{2m} = A.$$

Therefore $A + B + 1 = \left[\left(\frac{3B}{4}\right)^2 + \tfrac{1}{2}B\right] + B + 1$

$$= \left(\frac{3B}{4}\right)^2 + \tfrac{3}{2}B + 1$$

$$= \left(\frac{3B}{4} + 1\right)^2$$

$$= (\underbrace{33\ldots34}_{m})^2.$$

*AMM, 1895, p. 367, Problem 30, proposed by Cooper D. Schmitt, University of Tennessee, solved by George Zerr, Texarkansas College, Arkansas-Texas, and by H. C. Wilkes, Skull Run, West Virginia.

GROUPING THE NATURAL NUMBERS*

Suppose the natural numbers are divided into groups as shown:

$$(1), (2,3), (4,5,6), (7,8,9,10), (11,12,13,14,15),\ldots,$$

and that every second group is then deleted. Prove that the sum of the first k groups which remain is always k^4. For example, for $k=3$, we have

$$1+(4+5+6)+(11+12+13+14+15)=81=3^4.$$

Solution:

There are $n-1$ groups preceding the nth one, and they contain the first

$$1+2+3+\cdots+(n-1)=\frac{(n-1)n}{2} \text{ numbers.}$$

Therefore the first number in the nth group is $(n-1)n/2+1$. Similarly, the last number in the nth group is $n(n+1)/2$. Consequently, the sum $S(n)$ of the numbers in the nth group is

$$S(n)=\frac{n}{2}\left[\frac{(n-1)n}{2}+1+\frac{n(n+1)}{2}\right]=\frac{n(n^2+1)}{2}.$$

Now, the kth group that remains after the deletions is the $(2k-1)$th group in the original sequence. We want to prove, then, that

$$S(1)+S(3)+\cdots+S(2k-1)=k^4,$$

and we use induction, noting that $S(1)=1$.

*Scripta Mathematica, 1939, p. 218, Curiosa 56, by Dov Juzuk.

If $S(1) + S(3) + \cdots + S(2k-1) = k^4$, then

$$S(1) + S(3) + \cdots + S(2k+1) = k^4 + S(2k+1)$$

$$= k^4 + \frac{(2k+1)\left[(2k+1)^2+1\right]}{2},$$

which reduces to $(k+1)^4$.

TRIANGLES WITH SIDES IN ARITHMETIC PROGRESSION*

Prove that if the lengths of the sides of a triangle are in arithmetic progression, then the line joining the centroid to the incenter is parallel to one of the sides.

Solution:

Suppose the triangle is ABC, with sides $c < b < a$ (a is the side opposite vertex A, and so on). Then, because the sides are in arithmetic progression, we have $a + c = 2b$.

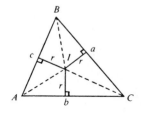 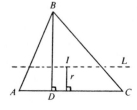 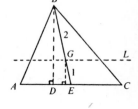

FIG. 93.

Let I denote the incenter and r the inradius (FIG. 93). Then the area is

$$\triangle ABC = \triangle IAC + \triangle IBC + \triangle IAB$$
$$= \tfrac{1}{2}br + \tfrac{1}{2}ar + \tfrac{1}{2}cr$$
$$= \tfrac{1}{2}r(a + b + c).$$

*AMM, 1940, p. 708, proposed by J. H. Butchart, Phillips University, solved by D. L. MacKay, Evander Hills High School, New York.

But the area of $\triangle ABC$ is $\frac{1}{2}BD \cdot b$, where BD is the altitude from B. Hence $\frac{1}{2}BD \cdot b = \frac{1}{2}r(a+b+c)$, and we have

$$\frac{r}{BD} = \frac{b}{a+b+c} = \frac{b}{3b} = \frac{1}{3}.$$

That is to say, I lies on a line L which is parallel to AC and which is one-third the way from AC to B. But the centroid G divides median BE in the ratio $2:1$, implying (by similar triangles) that G, too, lies on L. Thus GI determines L and therefore constitutes a line parallel to side AC.

FRACTIONS BY PERMUTATION*

Let $b_1, b_2, \ldots, b_n$ be any rearrangement of the positive real numbers $a_1, a_2, \ldots, a_n$. Prove that

$$\frac{a_1}{b_1} + \frac{a_2}{b_2} + \cdots + \frac{a_n}{b_n} \geqslant n.$$

Solution:

Since the arithmetic mean of n positive numbers is at least as great as their geometric mean, we have

$$\frac{1}{n}\left[\frac{a_1}{b_1} + \frac{a_2}{b_2} + \cdots + \frac{a_n}{b_n}\right] \geqslant \left[\frac{a_1}{b_1} \cdot \frac{a_2}{b_2} \cdots \frac{a_n}{b_n}\right]^{\frac{1}{n}} = 1,$$

from which the conclusion follows immediately.

*AMM, 1962, p. 59, Problem E1468, proposed by B. H. Bissinger, Lebanon Valley College, solved by Julius Vogel, Prudential Insurance Company.

ON BINOMIAL COEFFICIENTS*

What is the greatest common divisor g of the numbers

$$\binom{2n}{1}, \binom{2n}{3}, \binom{2n}{5}, \ldots, \binom{2n}{2n-1}?$$

Solution:

Any common divisor of a set of numbers divides their sum. The key here is to note that

$$\binom{2n}{1} + \binom{2n}{3} + \cdots + \binom{2n}{2n-1} = 2^{2n-1}.$$

We may deduce this as follows. The binomial theorem gives

$$(1+x)^{2n} = \binom{2n}{0} + \binom{2n}{1}x + \binom{2n}{2}x^2 + \cdots + \binom{2n}{2n}x^{2n}.$$

For $x = 1$, we see that the sum of all the coefficients is 2^{2n}. For $x = -1$, we see that the sum of these coefficients $\binom{2n}{r}$ for which r is odd is the same as the sum of those for which r is even. Thus the sum of the set of given numbers is one-half the grand total, namely 2^{2n-1}. The consequence of this is that the required greatest common divisor g must divide 2^{2n-1}, and therefore must be a power of 2.

Suppose that 2^k is the greatest power of 2 which divides n. Then $n = 2^k q$, where q is an odd number, and we have $\binom{2n}{1} = 2n = 2^{k+1}q$.

---

*AMM, 1971, p. 201, Problem E2227, proposed by N. S. Mendelsohn, University of Manitoba, solved by St. Olaf College Students.

Since g divides $\binom{2n}{1}$ and g is a power of 2, g cannot exceed 2^{k+1}. We shall show that $g = 2^{k+1}$ by deducing that 2^{k+1} does divide all the coefficients $\binom{2n}{r}$, $r = 1, 3, 5, \ldots, 2n - 1$.

Observe that

$$\binom{2n}{r} = \frac{(2n)!}{(2n-r)!r!} = \frac{2n}{r}\left[\frac{(2n-1)!}{(2n-r)!(r-1)!}\right] = \frac{2n}{r}\binom{2n-1}{r-1}$$

$$= \frac{2^{k+1}q}{r}\binom{2n-1}{r-1}.$$

Since r is odd and $\binom{2n}{r}$ is an integer, the factor 2^{k+1} in the numerator must survive the simplification of this expression and the conclusion follows.

THE FERMAT NUMBER F_{73}*

The numbers $F_n = 2^{(2^n)} + 1$, $n = 0, 1, 2, \ldots$, are called the Fermat numbers after the outstanding French mathematician Pierre de Fermat (1601–1665). The first five of these numbers are

$$3, 5, 17, 257, 65537, 4294967297.$$

You can imagine the tremendous size of $F_{73} = 2^{(2^{73})} + 1$. This problem raises the question of whether all the books in all the libraries in the whole world would contain enough room to record the ordinary decimal expression for this giant number F_{73}. In answering this question, we may use the very generous estimates concerning our books and libraries:

> 1 million libraries, each with 1 million books,
> each of 1000 pages, each containing 100 lines
> which will hold 100 digits apiece.

As a second part of the problem, we are asked for the last 3 digits that would be written down in recording F_{73}.

Solution:

1. For the given figures, the total capacity of the world's libraries would be

$$(100)(100)(1000)(1000000)(1000000) = 10^{19} \text{ digits.}$$

That's a lot of digits. Clearly, our problem is to get a line on the

---

*AMM, 1968, page 1119, Problem E2024, proposed by R. B. Eggleton, Avondale College, Cooranbong, Australia, solved by Harry Ploss, Cooper Union, New York City.

number of digits in F_{73}. Since

$$2^{10} = 1024 > 10^3,$$

we have

$$2^{73} = 8 \cdot 2^{70} > 8 \cdot 10^{21},$$

and

$$2^{(2^{73})} > 2^{8 \cdot 10^{21}} = (2^{80})^{10^{20}} = \left[(2^{10})^8 \right]^{10^{20}} > 10^{24 \cdot 10^{20}}.$$

Therefore F_{73} contains more than $24 \cdot 10^{20} = 240(10^{19})$ digits, requiring more than 240 library systems like ours in order to be recorded.

2. In order to find the last 3 digits of F_{73} we shall use, without proof, the following two remarkable discoveries:

(i) the square of a natural number and its 22nd power always end in the same two digits:
$$n^{22} \equiv n^2 (\text{mod } 100);$$

(ii) the cube of a natural number and its 103rd power always end in the same three digits:
$$n^{103} \equiv n^3 (\text{mod } 1000).$$

For nonnegative k, property (i) implies (mod 100) that
$$n^{k+22} = n^k \cdot n^{22} \equiv n^k \cdot n^2 = n^{k+2}.$$

Thus 20 may be subtracted from an exponent $\geqslant 22$ without altering the power's remainder (mod 100). Repeated applications permit the reduction by a multiple of 20 so long as the resulting exponent remains $\geqslant 2$. Similarly, property (ii) permits (mod 1000) the reduction of an exponent by a multiple of 100, so long as the resulting exponent remains $\geqslant 3$.

Accordingly, we have

$$2^{73} = 2^{60+13} \equiv 2^{13} (\text{mod } 100).$$

Thus (mod 100) we have

$$2^{73} \equiv 2^{13} = 2^3 \cdot 2^{10} = 8(1024) \equiv 8(24) = 192 \equiv 92.$$

Hence, for some integer q, we have

$$2^{73} = 100q + 92.$$

Using property (ii), we obtain

$$2^{(2^{73})} = 2^{100q + 92} \equiv 2^{92} (\text{mod } 1000).$$

Only simple arithmetic remains in calculating $2^{92} \equiv 896$ (mod 1000). Thus

$$F_{73} = 2^{(2^{73})} + 1 \text{ ends in 897.}$$

It was noted by St. Germain and Steen, who programmed a computer for the purpose, that the last 40 digits of F_{73} are

8947301518995672165296243935786246864897.

One of the most remarkable achievements of modern arithmetic is the classification as a composite number of the stupendously large

$$F_{1945} = 2^{(2^{1945})} + 1.$$

F_{73} is microscopic beside this colossus. With the above approach, however, it is even easier to determine the last 3 digits of F_{1945} than of F_{73}. Perhaps the reader would enjoy showing that F_{1945} ends in 297.

3. We conclude this section with a fully-supported alternative derivation of the last 3 digits of F_{73}.

Clearly

$$2^{10} = 1024 = 25t - 1 \text{ (for } t = 41) \equiv -1 \text{ (mod 25).}$$

The binomial theorem then gives

$$2^{100} = (2^{10})^{10} = (25t - 1)^{10} = (25t)^{10} - 10 \cdot (25t)^9 + \cdots$$
$$- 10 \cdot (25t) + 1.$$

Since each term, except the last, in this expansion is divisible by 125, we obtain

$$2^{100} \equiv 1 \text{ (mod 125).}$$

Now

$$2^{73} = (2^{10})^7 \cdot 2^3 \equiv (-1)^7 \cdot 2^3 (\text{mod } 25) \equiv -8 (\text{mod } 25),$$

giving $2^{73} = 25k - 8$ for some integer k. Also, it is obvious that $4|2^{73}$, giving

$$2^{73} \equiv 0 \equiv -8 \;(\text{mod } 4), \text{ and } 2^{73} = 4r - 8,$$

for some integer r. Then

$$25k - 8 = 4r - 8, \; 25k = 4r, \text{ implying } 4|k.$$

Letting $k = 4k_1$, we get

$$2^{73} = 100k_1 - 8 \equiv -8 \equiv 92 \;(\text{mod } 100),$$

and, for some integer q, we have

$$2^{73} = 100q + 92.$$

Thus

$$2^{(2^{73})} = 2^{100q+92} = 2^{92}(2^{100})^q \equiv 2^{92}(1)^q (\text{mod } 125)$$

and

$$2^{(2^{73})} \equiv 2^{92}(\text{mod } 125).$$

Let $2^{92} \equiv x \;(\text{mod } 125)$. Then

$$2^8 x \equiv 2^{100} \equiv 1 \;(\text{mod } 125),$$

and we have

$$2^8 x = 256x \equiv 6x \;(\text{mod } 125), \text{ and } 6x \equiv 1 \;(\text{mod } 125).$$

This gives

$$6x \equiv 126 \;(\text{mod } 125) \text{ and } x \equiv 21 \equiv -104 \;(\text{mod } 125).$$

Consequently,

$$2^{(2^{73})} \equiv x \equiv -104 \;(\text{mod } 125).$$

However, it is obvious that $8|2^{(2^{73})}$, implying

$$2^{(2^{73})} \equiv 0 \equiv -104 \;(\text{mod } 8).$$

Thus

$$2^{(2^{73})} = 125s - 104 = 8w - 104,$$

leading to $8|s$, $1000|125s$ and

$$2^{(2^{73})} = 1000v - 104 \equiv -104 \;(\text{mod } 1000) \equiv 896 \;(\text{mod } 1000).$$

Thus F_{73} ends in 897.

A CYCLIC QUADRILATERAL*

ABCD is a cyclic quadrilateral with opposite sides extended to meet at P and Q (Fig. 94). Prove that the quadrilateral EFGH determined on ABCD by the bisectors of the angles P and Q is always a rhombus.

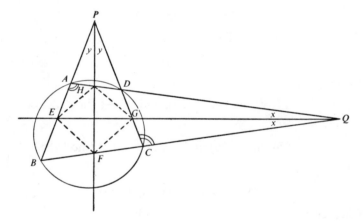

Fig. 94.

Solution:

Because ABCD is cyclic, the exterior angle DCQ is equal to the interior angle at the opposite vertex, A. Since QE bisects ∠ Q, the

*AMM, 1898, p. 143, Problem 90, proposed by George Zerr, Russell College, Lebanon, Virginia.

angles of $\triangle AQE$ are equal respectively to the angles of $\triangle CQG$. Hence

$$\angle CGQ = \angle AEQ.$$

But

$$\angle CGQ = \angle PGE \text{ (vertically opposite).}$$

Therefore

$$\angle PEG = \angle PGE, \text{ and } \triangle PEG \text{ is isosceles.}$$

Accordingly, the bisector of $\angle P$ is the perpendicular bisector of the base EG. Thus H and F, on this perpendicular bisector, are equidistant from E and G. Similarly E and G are equidistant from H and F, and $EFGH$ is a rhombus.

SPECIAL TRIPLES OF NATURAL NUMBERS*

Find all triples of 3 different natural numbers x, y, z, relatively prime in pairs, such that the sum of any two is divisible by the third.

Solution:

Suppose the numbers of such a triple are labeled so that $x < y < z$. Then

$$x + y < z + z = 2z,$$

implying that z cannot divide into the sum $x + y$ even as often as twice. Therefore it must go in once and we have $x + y = z$.

In this case, the sum $x + z = 2x + y$. Since this is divisible by y, we have $2x$ divisible by y. But $2x < 2y$, implying that y cannot divide into $2x$ as often as 2 times. Thus, as before, y must equal $2x$, and we have

$$x = x, y = 2x, z = x + y = 3x.$$

Since the numbers are to be relatively prime in pairs, x must be 1 and we obtain the unique answer $(1, 2, 3)$.

*AMM, 1957, p. 275, Problem E1234, proposed by Leo Moser and J. R. Pounder, University of Alberta, solved by E. P. Starke, Rutgers University.

THE SUMS OF THE PRIMES*

Let S_n denote the sum of the first n prime numbers:

$$s_n = 2 + 3 + 5 + \cdots + p_n.$$

Prove that between S_n and S_{n+1} there is always a perfect square.

Solution:

The claim is easily verified for $n = 1, 2, 3,$ and 4. Consider $n \geqslant 5$. Now, if the square roots of two positive real numbers x and y $(x < y)$ straddle an integer m, the perfect square m^2 will lie between x and y:

If $\sqrt{x} < m < \sqrt{y}$, then $x < m^2 < y$.

If $\sqrt{y} - \sqrt{x} > 1$, the interval between $\sqrt{x}$ and $\sqrt{y}$ is too large to avoid containing an integer, no matter where it might occur on the number scale. This is equivalent to

$$\sqrt{y} > 1 + \sqrt{x},$$
$$y > 1 + 2\sqrt{x} + x,$$
$$y - x > 1 + 2\sqrt{x}.$$

Thus we obtain the desired conclusion by showing that, for all n,

$$S_{n+1} - S_n > 1 + 2\sqrt{S_n}.$$

*AMM, 1969, p. 1151, Problem E2164, proposed by R. S. Luthar, University of Wisconsin at Waukesha, solved by Ivan Niven, University of Oregon (unpublished solution).

We have

$$S_n = 2 + 3 + 5 + 7 + 11 + \cdots + p_n.$$

For $n \geqslant 5$, this takes in at least $2 + 3 + 5 + 7 + 11$, from which the number 9 is missing. If we drop the 2 and include all the missing odd numbers, we gain at least $1 + 9 - 2 = 8$, and we obtain

$$S_n < 1 + 3 + 5 + \cdots + p_n,$$

the sum of all odd numbers from 1 to p_n.

Now, the sum of all the odd numbers up to $2k - 1$ is

$$1 + 3 + \cdots + (2k - 1) = \frac{k}{2} \left[1 + (2k - 1) \right] = k^2.$$

For $p_n = 2k - 1$, we get $k = \frac{1}{2}(1 + p_n)$. Therefore the sum of all the odd numbers up to p_n is $\frac{1}{4}(1 + p_n)^2$. Accordingly,

$$S_n < \tfrac{1}{4}(1 + p_n)^2$$

and

$$\sqrt{S_n} < \tfrac{1}{2}(1 + p_n),$$

or

$$2\sqrt{S_n} < 1 + p_n.$$

Observing that p_{n+1} must be at least $p_n + 2$ (since p_n and p_{n+1} cannot be consecutive), we have

$$
\begin{aligned}
S_{n+1} - S_n = p_{n+1} \\
\geqslant p_n + 2 \\
= 1 + (1 + p_n) \\
> 1 + 2\sqrt{S_n},
\end{aligned}
$$

as we set out to prove.

ANOTHER CURIOUS SEQUENCE*

This problem concerns a curious way of generating a sequence of natural numbers. Beginning with 2520 say, we obtain

$$2520, \quad 25, \quad 11, \quad 12, \quad 8, \quad 7, \quad 8, \dots.$$

The terms are determined by adding 1 to the sum of the prime divisors of the previous term, each prime being taken as often as is indicated by its exponent in the prime decomposition of the term. We have

$$2520 = 2^3 \cdot 3^2 \cdot 5 \cdot 7,$$

giving the second term to be

$$1 + 3 \cdot 2 + 2 \cdot 3 + 5 + 7 = 25;$$

then $25 = 5^2$ leads to the third term $1 + 2 \cdot 5 = 11$, and so on.

If n has the prime decomposition $n = p_1^{a_1} p_2^{a_2} \cdots p_k^{a_k}$, then the term following n, denoted $f(n)$, is $f(n) = 1 + a_1 p_1 + a_2 p_2 + \cdots + a_k p_k$.

Since $f(7) = 8$ and $f(8) = 7$, we see that an endless oscillation 8, 7, 8, 7,..., sets in as soon as an 8 or a 7 is encountered. Our problem is to show that sequences with initial terms $n > 6$ always contain a 7 or an 8 and thus, from some point onward, indefinitely repeat 8, 7, 8, 7,....

Solution:

In the example given in the statement of the problem, the sequence drops from 2520 to a mere 25 in the first step, and then

*AMM, 1973, p. 810, Problem E2356, proposed by J. B. Roberts, Reed College, solved by Hans Kappus, Switzerland.

down to the still lesser 11 in the next step. It is tempting to think that $f(n)$ is always less than n. However, 12 follows the 11, showing that this is not the case. In fact, for a prime p we have $f(p)=p+1$. Nevertheless, we are right in thinking that $f(n)$ is generally less than n. We shall see that the sequence increases only after terms which are primes, and then only by an amount 1, and that the rest of the time, except for $n=8$, the sequence drops at each step by at least 2:

For n composite, $f(n) \leqslant n-2$, except for $n=8$,

in which case the drop is only 1.

The conclusion follows readily from this central result.

At some point in the argument we must show that, for $n>6$, we always have $f(n)>6$. This follows from the examination of a few simple cases.

(i) If n has a prime divisor as great as 7, then $f(n)$, being greater than any prime divisor of n, exceeds 6. Thus we need consider only numbers whose prime divisors are 2, 3, and 5:

$$n=2^a \cdot 3^b \cdot 5^c.$$

(ii) If $c \geqslant 2$, then $f(n) \geqslant 1+2\cdot 5>6$.
If $c=1$, then, in order to have $n>6$, at least one of a or b must be nonzero, implying $f(n) \geqslant 1+5+2>6$. This leaves only numbers of the form

$$n=2^a \cdot 3^b, \text{ corresponding to } c=0.$$

(iii) If $b \geqslant 2$, then $f(n) \geqslant 1+2\cdot 3>6$. If $b=1$, then $n>6$ necessitates $a \geqslant 2$, and we have $f(n) \geqslant 1+2\cdot 2+3>6$. It remains, then, only to consider numbers of the form $n=2^a$.

(iv) For $n>6$, we must have $a \geqslant 3$, and we get $f(n) \geqslant 1+3\cdot 2>6$.

At this point, we are able to see how the overall argument goes. For n in the range of numbers >6, $f(n)$ remains in this range. A prime p in this range is odd, making $f(p)=p+1$ an even number and therefore composite. Consequently, pairs of consecutive terms in the sequence can only be of the following three types (we cannot have a prime followed by another prime):

$$\ldots\;,\text{prime, composite, }\ldots$$

$$\ldots\;,\text{composite, prime, }\ldots$$

$$\ldots\;,\text{composite, composite, }\ldots\;.$$

If either term of a pair is 8, then the oscillation is already in progress and the sequence accords with our hypothesis. On the basis that $f(n)$ drops by at least 2 for n a composite number $\neq 8$, we see from the three types of pairs that, over two consecutive steps, every other situation leads to a net drop of at least 1:

For example,

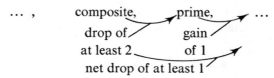

Thus the alternate terms of the sequence give us a subsequence which, except for the appearance of an 8, is strictly decreasing. Yet all of these terms remain >6. Consequently, they eventually must hit bottom. Generally this means that the number 7 is reached and the oscillation takes over in the full sequence. The only way to avoid this is to suspend the necessity for a drop of at least 1 over a pair of terms by the occurrence of a term equal to the exceptional value 8. However, this, too, provides the oscillation and the proof is complete. It remains for us to establish the fundamental result:

For n composite, $n>6, n\neq 8, f(n)\leqslant n-2,$

that is, for composite $n\geqslant 9, f(n)\leqslant n-2.$

We proceed by induction. Since $f(9)=7$, the property holds for the number 9. Let n denote a composite natural number greater than 9, and let us take as induction hypothesis that the property holds for all lesser composite natural numbers which are also $\geqslant 9$. We show that the property is then inherited by the number n, itself.

Since n is composite, it has at least one nontrivial factorization

$$n = k_1 k_2, \text{ where } 1 < k_1, k_2 < n.$$

If either of these factors k_i is a prime number, then $f(k_i) = k_i + 1$. If k_i is composite and at least 9, then it falls under the scope of the induction hypothesis, and we have $f(k_i) \leqslant k_i - 2$. Otherwise, k_i is either 4, 6, or 8, and $f(k_i)$ is, respectively, 5, 6, or 7. Whatever be the case, we see that $f(k_i)$ is never greater than $k_i + 1$. Thus

$$f(k_1) \leqslant k_1 + 1 \qquad \text{and} \qquad f(k_2) \leqslant k_2 + 1.$$

Suppose that

$$n = p_1^{a_1} p_2^{a_2} \cdots p_k^{a_k}.$$

The factorization $n = k_1 k_2$ is effected by partitioning the prime divisors of n into two nonempty groups:

$$n = \underbrace{\left(p_1^{b_1} p_2^{b_2} \cdots p_k^{b_k} \right)}_{k_1} \underbrace{\left(p_1^{a_1 - b_1} p_2^{a_2 - b_2} \cdots p_k^{a_k - b_k} \right)}_{k_2}.$$

In general we have

$$f(t) = 1 + \text{(the sum of the prime divisors of } t, \text{ each taken}$$

$$\text{appropriately often)}.$$

Accordingly, the sum of the prime divisors of k_1, with the appropriate repetitions, is $f(k_1) - 1$. Similarly, for k_2 and n, we have, respectively, $f(k_2) - 1$ and $f(n) - 1$. However, because $n = k_1 k_2$, we have that

(the sum of the prime divisors of n, taken appropriately often)

= (the sum of the prime divisors of k_1, taken appropriately often)

+ (the sum of the prime divisors of k_2, taken appropriately often).

This gives

$$f(n) - 1 = f(k_1) - 1 + f(k_2) - 1,$$

and

$$f(n) = f(k_1) + f(k_2) - 1.$$

Because $f(k_1) \leqslant k_1 + 1$, and $f(k_2) \leqslant k_2 + 1$, we obtain

$$f(n) \leqslant k_1 + 1 + k_2 + 1 - 1,$$

and

$$\boxed{f(n) \leqslant k_1 + k_2 + 1}.$$

Now, we know that each of k_1 and k_2 is at least 2 and that their product n is at least 9. Consequently, we are able to deduce easily that the product

$$(k_1 - 1)(k_2 - 1) \text{ is at least 4:}$$

(i) If one of k_1, k_2 is the minimum value 2, making its $k_i - 1 = 1$, then, because $n = k_1 k_2 \geqslant 9$, the other $k_i \geqslant 5$. In this case, $(k_1 - 1)(k_2 - 1) \geqslant 1 \cdot 4 = 4$.

(ii) Otherwise, each $k_i \geqslant 3$, making each $k_i - 1 \geqslant 2$, and their product $\geqslant 4$.

The proof concludes with the clever observation that $k_1 + k_2 + 1$ and $(k_1 - 1)(k_2 - 1)$ are nicely related. We have

$$(k_1 - 1)(k_2 - 1) = k_1 k_2 - k_1 - k_2 + 1$$
$$= k_1 k_2 - (k_1 + k_2 + 1) + 2.$$

Thus

$$k_1 + k_2 + 1 = k_1 k_2 + 2 - (k_1 - 1)(k_2 - 1)$$
$$= n + 2 - (k_1 - 1)(k_2 - 1).$$

By the inequality above, then, we have

$$f(n) \leqslant n + 2 - (k_1 - 1)(k_2 - 1).$$

Since $(k_1 - 1)(k_2 - 1) \geqslant 4$, we obtain

$$f(n) \leqslant n + 2 - 4$$

and

$$f(n) \leqslant n - 2.$$

THE ELLIPSE AND THE LATTICE*

No matter how a circle of radius 5 is tossed onto a (unit, square) lattice, it is far too big to avoid covering some lattice point. (We consider our figures to contain their boundaries, that is, to be closed.) Of course, a very small circle might well miss all the lattice points. Since every point of the plane is at most $\sqrt{2}/2$ from a lattice point ($\sqrt{2}/2$ is one-half the diagonal of the fundamental unit-square cell of the lattice), no matter where the center of a circle of radius $\geqslant \sqrt{2}/2$ happens to land, the circle will extend far enough to cover at least one lattice point. It is more of a challenge to prove that an $a \times b$ rectangle, tossed at random onto a lattice, can be guaranteed to cover at least one lattice point if and only if $a \geqslant 1$ and $b \geqslant \sqrt{2}$. And it is a complicated matter to show that a triangle can always be counted on to cover at least one lattice point if and only if its area $\geqslant c^2/2(c-1)$, where c is the length of its longest side. However, the case of an ellipse is a most interesting topic.

Ellipses come in an infinity of shapes. Nevertheless, the condition governing whether or not an ellipse always covers a lattice point is simply that, when in standard position (that is, center at the origin, axes along the coordinate axes), it covers the point $(1/2, 1/2)$. Prove this theorem.

*AMM, 1967, pp. 353-362, "Lattice Point Coverings by Plane Figures", by Ivan Niven, University of Oregon, and H. S. Zuckerman, University of Washington; corrections p. 952.

Solution:

Let the equation of the given ellipse be $b^2x^2 + a^2y^2 = a^2b^2$. The condition that it cover $(1/2, 1/2)$ is given by

$$\tfrac{1}{4}b^2 + \tfrac{1}{4}a^2 \leqslant a^2b^2, \qquad \text{or} \qquad a^2 + b^2 \leqslant 4a^2b^2.$$

The theorem claims, then, that the ellipse will always cover at least one lattice point if and only if $a^2 + b^2 \leqslant 4a^2b^2$.

(a) *Sufficiency*:

Suppose that $a^2 + b^2 \leqslant 4a^2b^2$. Consider now a (unit, square) lattice L of points (u, v) determined by given u and v axes. Let our ellipse E be tossed on L to land where it will (FIG. 95). It is our desire to work analytically with reference to the axes of E as the x and y axes of coordinates. Relative to the xy-reference frame, E is in standard position and has equation $b^2x^2 + a^2y^2 = a^2b^2$. The lattice lines of L may cut across our xy-plane at any angle. Since the perpendicular axes u and v point in all four directions of the compass, one of the axes must have a slope m, relative to the xy-system, in the range $-1 < m \leqslant 1$. Suppose the u-axis has such a slope m. The lattice lines $v = n$ in L, n an integer, constitute a set of equally spaced parallel lines of slope m in the xy-system, and they intersect the y-axis in a set of equally-spaced points B_n. The first thing we determine is the distance B_nB_{n+1} between consecutive intersections on the y-axis.

On the line T, of slope m, through the origin O, let a point P be taken to have x-coordinate 1 (FIG. 96). Let the ordinate to P be PN. Since the slope $PN/ON = m$, we have $PN = m$, and $OP = \sqrt{1 + m^2}$. Now, drop the perpendicular $B_{n+1}S$ to $v = n$. Since corresponding angles $B_{n+1}B_nS$ and B_nOP are equal, we have equal complements in angles $SB_{n+1}B_n$ and PON. Consequently, the triangles $B_{n+1}B_nS$ and PON are congruent ($B_{n+1}S = ON = 1$), and the required

$$B_{n+1}B_n = OP = \sqrt{1 + m^2} \; .$$

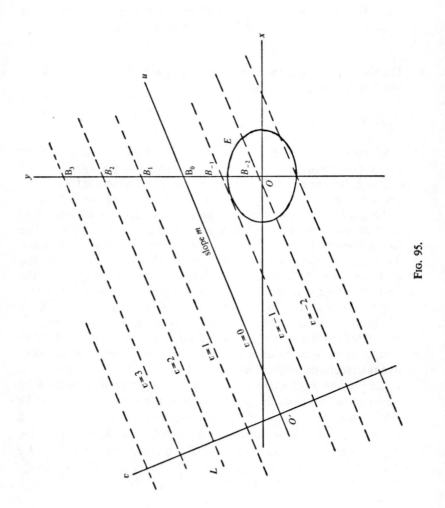

Fig. 95.

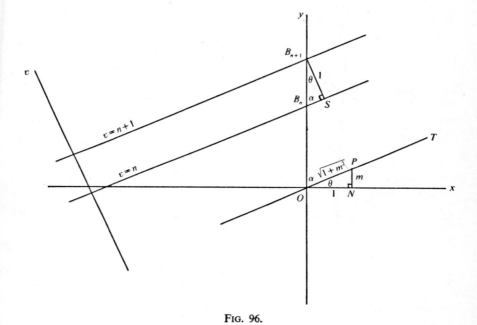

FIG. 96.

The points B_n, then, march down the y-axis with a step of size $\sqrt{1+m^2}$. Therefore some $B_n = (0,y)$ must occur with y in the range $-\frac{1}{2}\sqrt{1+m^2} < y \leqslant \frac{1}{2}\sqrt{1+m^2}$, obtained by centering a step at the origin (it is too big a range for the B_n to step over completely) (FIG. 97). If B_n has coordinates $(0,k)$, the equation of the line L_n of lattice points ($v = n$ in L) which it yields is

$$y = mx + k, \text{ where } -\frac{1}{2}\sqrt{1+m^2} < k \leqslant \frac{1}{2}\sqrt{1+m^2}.$$

Now, at this point, there is no guarantee that L_n will intersect our ellipse E. However, solving their equations, we find

$$y = mx + k,$$
$$b^2x^2 + a^2y^2 = a^2b^2,$$
$$b^2x^2 + a^2(mx + k)^2 = a^2b^2,$$

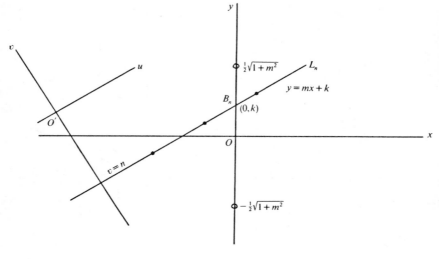

FIG. 97.

which reduces to

$$x = \frac{-mka^2 \pm ab\sqrt{b^2 + a^2m^2 - k^2}}{b^2 + a^2m^2}.$$

Since E covers the point $(1/2, 1/2)$, we have $a^2 + b^2 \leqslant 4a^2b^2$, and therefore $(1/b^2) + (1/a^2) \leqslant 4$, giving $1/b^2 < 4$ and $1/a^2 < 4$, since a and b are nonzero. Thus we have $b^2 > \frac{1}{4}$ and $a^2 > \frac{1}{4}$. As a result, we have

$$k^2 \leqslant \left(\frac{1}{2}\sqrt{1 + m^2}\right)^2 = \frac{1}{4}(1 + m^2) = \frac{1}{4} + \frac{m^2}{4} < b^2 + a^2m^2.$$

Thus the quantity under the root sign, namely $b^2 + a^2m^2 - k^2$, is positive, implying that L_n does indeed cross our ellipse E.

Now we pursue the length d of the intercept that E cuts from L_n. If we are able to show that d is at least 1, then the lattice points of L_n, marching along at unit intervals, must have at least one fall in E.

Let x_1 and x_2 denote the x-coordinates of the points of intersection of E and L_n. Since L_n has xy-equation $y = mx + k$, the points of intersection are $(x_1, mx_1 + k)$ and $(x_2, mx_2 + k)$. Thus

$$d^2 = (x_2 - x_1)^2 + (mx_2 - mx_1)^2 = (x_2 - x_1)^2 (1 + m^2).$$

Having solved earlier for the points of intersection, we see that the values of x_1 and x_2 are

$$\frac{-mka^2 \pm ab\sqrt{b^2 + a^2m^2 - k^2}}{b^2 + a^2m^2}.$$

The square of their difference, then, is

$$\left[\frac{2ab\sqrt{b^2 + a^2m^2 - k^2}}{b^2 + a^2m^2}\right]^2 = \frac{4a^2b^2(b^2 + a^2m^2 - k^2)}{(b^2 + a^2m^2)^2}.$$

Thus

$$d^2 = \frac{4a^2b^2(b^2 + a^2m^2 - k^2)(1 + m^2)}{(b^2 + a^2m^2)^2},$$

and

$$d^2(b^2 + a^2m^2)^2 = 4a^2b^2(b^2 + a^2m^2 - k^2)(1 + m^2).$$

Subtracting $(b^2 + a^2m^2)^2$ from each side gives

$$(d^2 - 1)(b^2 + a^2m^2)^2 = 4a^2b^2(b^2 + a^2m^2 - k^2)(1 + m^2) - (b^2 + a^2m^2)^2.$$

Upon showing that the right side of this equation is nonnegative, we conclude that $d^2 - 1 \geqslant 0$, and $d \geqslant 1$, as desired.

Now, $0 \leqslant (a - b)^2$ gives $2ab \leqslant a^2 + b^2$. Since we have

$$\boxed{a^2 + b^2 \leqslant 4a^2b^2},$$

it follows that $2ab \leqslant 4a^2b^2$, and we get $1 \leqslant 2ab$. Thus $1 \leqslant 2ab < a^2 + b^2$, and

$$\boxed{a^2 + b^2 - 1 \geqslant 0}.$$

Since we also have $k^2 \leqslant (1 + m^2)/4$, we get

$$4a^2b^2(b^2 + a^2m^2 - k^2)(1 + m^2) - (b^2 + a^2m^2)^2$$

$$\geqslant 4a^2b^2\left(b^2 + a^2m^2 - \frac{1 + m^2}{4}\right)(1 + m^2) - (b^2 + a^2m^2)^2$$

$$= (m^4a^2 + b^2)(4a^2b^2 - a^2 - b^2) + 4m^2a^2b^2(a^2 + b^2 - 1).$$

This last step is seen to be correct by multiplying out all the expressions. From the inequalities above, we have

$$4a^2b^2 - a^2 - b^2 \geqslant 0 \text{ and } a^2 + b^2 - 1 \geqslant 0.$$

Thus the result is nonnegative, and the sufficiency is established.

(b) *Necessity*:

Suppose, on the other hand, that E always covers a lattice point. It is a simple matter to see that it must cover the point $(1/2, 1/2)$ when in standard position.

Suppose, to the contrary, that it does not. Then translate the axes by shifting the origin to the point $(1/2, 1/2)$. The effect of this is to interchange the lattice points and the centers of the fundamental unit-square cells of the lattice (FIG. 98). If a region covers a lattice point (x, y), then, after the translation, it is moved

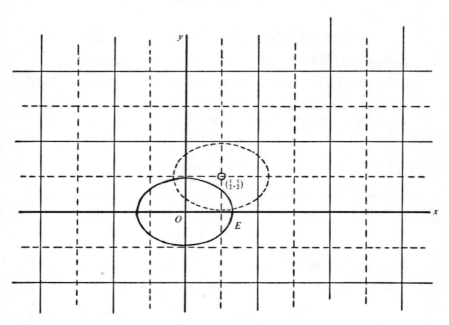

FIG. 98.

to cover a center, and vice-versa. If E fails to cover the center $(1/2, 1/2)$, then, because of its symmetry, it fails to cover the center of any lattice square. Accordingly, when shifted, it will fail to cover any lattice point, contradicting the given property that E always covers at least one lattice point in a (unit, square) lattice (in this case, the equal lattice of the centers). Thus E must cover the point $(1/2, 1/2)$ and the theorem follows.

ARCHIMEDES TRIANGLES*

One of the remarkable achievements of the great Archimedes is the determination of the area of a segment of a parabola (that is, the part cut off by a chord). He found that the segment on chord AB is $2/3$ the area of $\triangle PAB$ which is determined by the tangents at A and B. Such a triangle PAB is called an "Archimedes" triangle (FIG. 99).

It can be shown that the median PT to the chord of an Archimedes triangle is parallel to the axis of the parabola and that

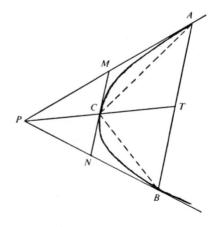

FIG. 99.

*AMM, 1935, p. 606, "Properties of Parabolas Inscribed in a Triangle", by J. A. Bullard, University of Vermont; see also AMM, 1937, p. 368.

it crosses the parabola at a point C at which the tangent MCN is parallel to the chord AB. It also turns out that M and N are the midpoints of PA and PB. Therefore the triangle CAB has the same base as $\triangle PAB$ but only one-half its altitude, making $\triangle CAB = \frac{1}{2} \cdot \triangle PAB$. Consequently,

$$\text{the parabolic segment} = \tfrac{2}{3} \triangle PAB = \tfrac{4}{3} \triangle CAB.$$

Now, a given triangle is an Archimedes triangle for three parabolas, each touching two of the sides at vertices. An Archimedes triangle with its three parabolas is a configuration which is rich in interesting properties (FIG. 100).

(i) The parabolas always cross each other on the medians of the triangle (e.g., NTM is a median).

(ii) These points of intersection always occur $1/9$ the way down a median (from the midpoint of the side)(e.g., $MT=(1/9)$ MN, or $MT:TN=1:8$).

(iii) The parabolas and the medians between them divide the triangle into 18 sections, 12 of which are bounded by two line segments and an arc, and 6 of which are bounded by two arcs and one segment. These 12 sections all have the same area (5/162 of

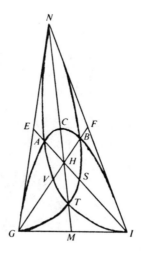

FIG. 100.

the area of the triangle), and the 6 sections also have the same area (17/162 of the triangle).

(iv) The tangents to the parabolas at a point of intersection (e.g., T) trisect the adjacent side (i.e., P, Q trisect GI) (FIG. 101).

(v) The nine-point circle (through the midpoints of the sides) intersects the median and altitude from a vertex (e.g., GF and GK) in two points (X and Y) which determine a line that is perpendicular to the median; this makes it perpendicular also to the axis of the (appropriate) parabola, and therefore parallel to the directrix. In fact this line (XY) *is* the directrix.

(vi) A line through the focus f parallel to the nontangential side of the triangle cuts from the median (GF) to that side a segment (ZF) which is equal to the distance along the median from the vertex (G) to the directrix (i.e., $GX = ZF$).

In this section we conclude our collection of stories with the consideration of properties (i), (ii), and half of (iii).

(i) and (ii). These properties are established together by showing that the point W which divides median MN in the ratio $1 : 8$ lies on each of the parabolas which are tangent to side GI (FIG. 102).

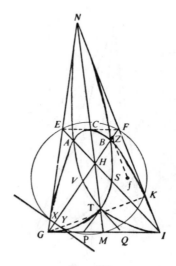

FIG. 101.

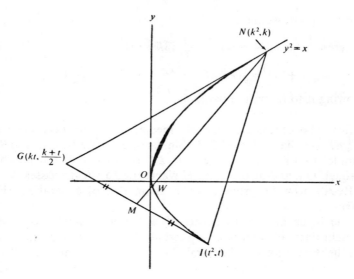

FIG. 102.

Since these parabolas give equivalent cases, we consider only the one through N and I. Suppose that its equation is $y^2 = x$ and that the coordinates of N and I, respectively, are (k^2, k) and (t^2, t). Then the tangents GN and GI have equations

$$x - 2ky + k^2 = 0 \text{ and } x - 2ty + t^2 = 0.$$

Solving gives G to be $(kt, (k+t)/2)$. (In passing, we note that the ordinate $(k+t)/2$ of G is the same as that of the midpoint of NI, showing the median to the chord NI is parallel to the axis of the parabola.)

The midpoint M of GI is $(t(k+t)/2, k+3t)/4$. The point which divides MN in the ratio $1:8$ is

$$\left[\frac{k^2 + 8\left(\dfrac{t(k+t)}{2} \right)}{9}, \frac{k + 8\left(\dfrac{k+3t}{4} \right)}{9} \right], \text{ or}$$

$$\left(\frac{1}{9}\left[k^2 + 4t(k+t) \right], \frac{1}{9}\left[k + 2(k+3t) \right] \right).$$

For this point we have

$$y^2 = \frac{1}{81}\left[k + 2(k + 3t)\right]^2 = \frac{1}{81}(3k + 6t)^2$$
$$= \tfrac{1}{9}(k + 2t)^2 = \tfrac{1}{9}(k^2 + 4kt + 4t^2) = \tfrac{1}{9}\left[k^2 + 4t(k + t)\right] = x,$$

showing it to lie on the parabola.

(iii). The centroid H trisects medians EI and FG. Since $EA = \tfrac{1}{9}EI$, we have $EA = \tfrac{1}{3}EH$. Similarly $FB = \tfrac{1}{3}FH$. Thus AB is parallel to EF. As noted above, EF is tangent to the parabola through G and I at the point C where median NM crosses it.

Recall now the property of a "diameter" of a parabola (Fig. 103):

The locus of the midpoints of a set of parallel chords is a straight line called a diameter, and all diameters of a parabola are parallel to the axis of the parabola; also, the tangent at the "end"

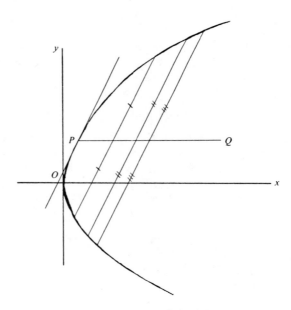

Fig. 103.

of a diameter is parallel to the set of chords it bisects. Therefore, a straight line PQ drawn from a point P on a parabola, parallel to the axis, bisects every chord which is parallel to the tangent at P.

Now, we have just seen that chord AB is parallel to tangent ECF, where CO, being part of median NM, is parallel to the axis of the parabola (FIG. 104). Accordingly, CO is a diameter, and CO bisects all the chords parallel to EF (and AB). We conclude, then, that CO bisects the *area* of segment ABC. AB is one of the chords that is bisected by CO. Therefore OH bisects $\triangle ABH$, and we see that CH bisects $AHBC$. Thus the area of $AHC = \frac{1}{2} AHBC = \frac{1}{2}$(segment $ABC + \triangle ABH$). We turn, then, to the calculation of segment ABC and $\triangle ABH$.

Because EF and AB are parallel, triangles EFH and ABH are similar. Therefore

$$\frac{\triangle ABH}{\triangle EFH} = \left(\frac{AH}{EH}\right)^2 = \left(\frac{2}{3}\right)^2 = \frac{4}{9},$$

and $\triangle ABH = \frac{4}{9} \triangle EFH$.

Now, EF is parallel to GI, making triangles EFH and HGI

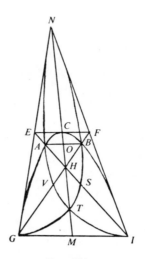

FIG. 104.

similar. Since the ratio of corresponding sides is $1/2$,

$$\triangle EFH = \tfrac{1}{4} \triangle HGI.$$

However, the centroid H trisects median MN, implying (by similar triangles) that the corresponding altitudes to GI in triangles HGI and NGI are also in the ratio of $1:3$. Hence $\triangle HGI$ is one-third $\triangle NGI$ and

$$\triangle EFH = \frac{1}{4}\left(\frac{1}{3}\triangle NGI\right) = \frac{1}{12}\triangle NGI.$$

Accordingly,

$$\triangle ABH = \frac{4}{9}\triangle EFH = \frac{4}{9}\left(\frac{1}{12}\triangle NGI\right) = \frac{1}{27}\triangle NGI.$$

Finally, segment $ABC = \tfrac{4}{3}\triangle ABC$. Because A and B divide EH and FH in the ratio of $1:2$, the distance between the parallel lines EF and AB, that is, the altitude to base AB of $\triangle ABC$, is one-half the altitude from H to AB in $\triangle ABH$. Therefore $\triangle ABC$ is one-half $\triangle ABH$ and we have

$$\triangle ABC = \frac{1}{2}\triangle ABH = \frac{1}{2}\left(\frac{1}{27}\triangle NGI\right) = \frac{1}{54}\triangle NGI.$$

Hence

$$\text{segment } ABC = \frac{4}{3}\triangle ABC = \frac{4}{3}\left(\frac{1}{54}\triangle NGI\right) = \frac{2}{81}\triangle NGI.$$

Finally, then,

$$\text{region } ACH = \frac{1}{2}\left(\frac{2}{81}\triangle NGI + \frac{1}{27}\triangle NGI\right)$$

$$= \left(\frac{1}{81} + \frac{1}{54}\right)\triangle NGI$$

$$= \frac{5}{162}\triangle NGI.$$

EXERCISES

A reference for the problem and its solution is given at the end of its statement.

1. In overtaking a freight, a passenger train, which is x times as fast, takes x times as long to pass as it takes the two trains to pass when going in opposite directions. Find x. (AMM, 1960, p. 475, problem E1386.)

2. From each of the centers of two given circles tangents are drawn to the other circle (FIG. 105). Prove that equal chords are intercepted on the circumferences. (AMM, 1933, p. 456.)

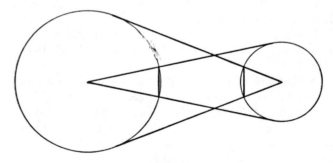

FIG. 105.

3. Find the least natural number having the property that the sum of its digits does *not* divide the sum of the cubes of its digits. (AMM, 1948, p. 579, problem E802.)

4. Let A, B, C denote any three points on a parabola which has vertical axis (i.e., parallel to the y-axis). Let m_A denote the slope of the tangent at A, and let m_{AB} denote the slope of the chord AB, etc. Prove the surprising property that

$$m_A = m_{AB} + m_{AC} - m_{BC}.$$

(AMM, 1965, p. 667, problem E1701.)

5. Prove that the "magic constant" of a 3×3 magic square is always a multiple of 3. (AMM, 1897, p. 189.)

6. Somebody received a check, calling for a certain amount in dollars and cents. When he cashed the check the teller made a mistake, paying him in dollars the amount written in cents, and vice-versa. Later, after spending $3.50, he suddenly realized that he had twice the amount called for by the check. What was the amount on the check? (AMM, 1941, p. 212, problem E430.)

7. Show that in every tetrahedron there must be at least one vertex at which each of the face angles is acute. (AMM, 1935, p. 453, problem E141.)

8. Given the polynomial $f(x)$ with integral coefficients and an odd number a and an even number b such that $f(a)$ and $f(b)$ are both odd. Prove that $f(x) = 0$ has no integral roots. (AMM, 1960, p. 760, no solution.)

9. At no vertex of the simple polyhedron P do exactly 3 edges meet. Prove that at least 8 of the faces of P are triangles. (AMM, 1951, p. 421, problem E945.)

10. Prove that every power a^n, a and n natural numbers, $n > 1$, is the sum of a consecutive odd numbers. (AMM, 1947, p. 165, problem E726.)

11. A chord of constant length slides around in a given circle. The endpoints of the chord are projected (orthogonally) upon a fixed diameter of the circle. The feet of the projections and the midpoint of the chord determine a triangle. Prove that this triangle is isosceles and never changes shape as the chord slides around. (AMM, 1936, p. 186, problem E171.)

12. A announces a two-digit number from 01 to 99. B reverses the digits of this number and adds to it the sum of its digits and then announces his result to A. A repeats this pattern and they continue taking turns doing so. All the numbers which result are reduced modulo 100, so that only two-digit numbers are announced. What choices has A for his initial call in order to insure that B will at some time announce 00? (AMM, 1949, p. 105, problem E816.)

13. Prove that no arrangement of the five even digits and also no arrangement of the five odd digits ever yields a perfect square. (AMM, 1937, p. 248, problem E232.)

14. A variable circle is drawn through two fixed points to cut a fixed circle in two points. Prove that all the common chords of the fixed circle and the variable circles are concurrent. (AMM, 1895, p. 17, problem 32.)

15. Find the maximum and minimum number of "Friday the 13th's" that can occur in a year. (AMM, 1963, p. 759, problem E1541.)

16. Find two natural numbers such that their sum will be a factor of their product. (AMM, 1961, p. 804, problem E1452.)

17. Given a circle, center O, and a point P not on the circle. A line is drawn through P cutting the circle in two points at which the tangents are drawn. These tangents meet the line through P which is perpendicular to OP at C and D. Prove P is the midpoint of CD. (Mathematics News Letter (forerunner of the Mathematics Magazine), 1933–34, p. 170, problem 52.)

18. Prove, for all $n = 1, 2, 3, \ldots,$ that

$$(1^5 + 2^5 + \cdots + n^5) + (1^7 + 2^7 + \cdots + n^7) = 2(1 + 2 + \cdots + n)^4.$$

(AMM, 1915, p. 99, problem 419.)

19. Prove that if perpendiculars are drawn from the feet of the altitudes of a triangle to each of the other two sides, the six feet of these perpendiculars lie on a circle. (This is known as the Taylor circle of the triangle.) (NMM, 1943–44, p. 40, problem 485.)

20. Show that a natural number is determined uniquely by a knowledge of the product of all its divisors. (MM, 1964, p. 57, problem 518.)

21. Prove that if, in a triangle, two medians are perpendicular, then the three medians are the sides of a right-angled triangle which has the third median as hypotenuse. (AMM, 1902, p. 164, problem 177.)

22. Let f_n be the nth term of the sequence defined by

$$f_n = -f_{n-1} - 2f_{n-2}, f_1 = 1, f_2 = -1.$$

Prove that $2^{n+1} - 7f_{n-1}^2$ is always a perfect square. (AMM, 1973, p. 696, problem E2367.)

23. Farmer Jones has a cart with square wheels. However, it suits his needs since he is able to use it to travel the washboard road without any bumping. Assuming no slipping of the wheels, describe the shape of the washboard road. (AMM, 1965, p. 82, problem E1668.)

24. Find all values of r such that no value of $n!$ ends in exactly r zeros. (MM, 1953, p. 54, problem 158.)

25. If a circle has its center on an equilateral hyperbola and passes through the point on the hyperbola which is diametrically opposite that point, then the remaining three points of intersection of the circle and the hyperbola always determine an equilateral triangle (FIG. 106). (NMM, 1943–44, p. 247, problem 531.)

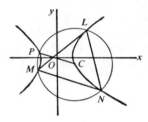

FIG. 106.

LIST OF THE PROBLEMS BY TOPIC

I. ALGEBRA, ARITHMETIC, NUMBER THEORY, SEQUENCES, PROBABILITY

II. COMBINATORICS, COMBINATORIAL GEOMETRY (MAXIMA AND MINIMA)

III. GEOMETRY (MAXIMA AND MINIMA)